Printing
Industry
Advisory
Committee

Fire safety in the printing industry

London: HMSO

First published 1992

ISBN 0 11 886375 4

Enquiries regarding this or any other Health and Safety Executive publications should be made to:

HSE Information Centre
Broad Lane
Sheffield S3 7HQ
Telephone: 0742 892345
Fax: 0742 892333

Front cover photograph courtesy of
Devon Fire and Rescue Service

Contents

APPENDICES

PREFACE

This booklet is the result of a joint initiative by the Health and Safety Executive, the Home Office, the Scottish Office and the Health and Safety Commission's Printing Industry Advisory Committee (PIAC) to provide guidance on appropriate fire precautions to those involved in the printing industry and allied trades so that all can work in safety. It has been prepared by a PIAC Working Party.

The purpose of this guidance is to create a greater awareness of the fire and explosion risks associated with the printing industry, to advise on appropriate basic fire precautions and staff training and to outline the relevant statutory responsibilities imposed on those concerned with the management of premises. An action plan for fire risk management is included.

It is an authoritative document that will be used by health and safety inspectors in describing reliable and fully acceptable methods of achieving health and safety in the workplace. It remains open to employers to achieve equivalent levels of health and safety using other acceptable means; if so, reference is likely to be made to this document by inspectors to demonstrate the level that has to be achieved. Equally, while it has no legal force, its standing as agreed practical guidance means that it may be referred to in a court or tribunal to demonstrate the standards that need to be met under the law.

INTRODUCTION

1 This guidance is aimed at managers who have responsibility for fire safety to help them fulfil their duties under the Fire Precautions Act 1971, the Health and Safety at Work etc Act 1974 (HSW Act) and other relevant legislation. It will also be useful to health and safety advisers, trainers, technical staff, supervisors, safety representatives and all others who work in the industry.

2 Paper, board, ink and the majority of other materials used in the printing industry are combustible. While solid blocks of paper may not be easily ignited, once they have caught fire flames can spread rapidly and be difficult to extinguish. Loose paper and wrapping and flammable liquids can ignite easily and spread the fire to other materials.

3 The fire service is called every year to several hundred fires in the printing industry; there have also been a number of explosions. In some incidents people have been killed or injured. In many others there has been extensive damage to buildings, equipment and materials; in addition to the immediate material damage there are the inevitable consequential losses, such as reduced turnover and dispersal of skilled staff. For summaries of some fires and explosions in the industry see Appendix 1.

4 The guidance does not deal with potential health hazards arising from some of the flammable materials used in the industry. These are considered in other publications from PIAC (see References 1, 2 and 3).

LEGAL REQUIREMENTS

5 The term 'fire precautions' includes matters which are the subject of legal requirements under the Fire Precautions Act 1971 (Ref 4) and the Health and Safety at Work etc Act 1974 (Ref 5). The advice in this booklet is supported by the Home Departments (the Home Office and the Scottish Office) and Health and Safety Executive (HSE) and is relevant whichever legislation applies.

6 The Fire Precautions Act 1971 is the responsibility of the Home Departments and is enforced by the fire authorities and, for Crown-occupied and Crown-owned premises by the Fire Service Inspectorates of the Home Departments. It deals with general fire precautions and includes the provision of means of escape, means for fighting fire, means for giving warning in case of fire and the training of staff in fire safety.

7 The HSW Act and relevant Regulations cover the provision of process fire precautions which are intended to prevent the outbreak of fire or mitigate the consequences should one occur. Matters falling within the scope of the Act include the storage of flammable materials, the control of flammable vapours, standards of housekeeping, safe systems of work, the control of sources of ignition and provision of appropriate training. In printing firms the Act will normally be enforced by HSE inspectors and in offices by inspectors from the local authority (see the telephone directory for the location of the local HSE office).

8 In England and Wales the licensing authority for the keeping of petroleum spirit or petroleum mixtures is generally the county council or in metropolitan counties and London, the fire and civil defence authority; in Scotland it is normally the local regional or islands council.

Fire Precautions Act 1971

9 The Fire Precautions Act 1971 as amended by the Fire Safety and Safety of Places of Sport Act 1987 is the principal instrument for fire safety in factories and associated offices and is designed to ensure the adequate provision of means of escape in case of fire and related fire precautions.

10 Inspectors enforcing the Fire Precautions Act have the power of entry into premises without prior notice and can issue notices requiring specified improvements. They may also serve a notice prohibiting or restricting the use of *any* premises (whether or not it requires a fire certificate) if the conditions are or will become, in the case of fire, such that they present a serious risk to people on the premises.

Premises requiring a fire certificate

11 A fire certificate is generally required if more than 20 people are at work in the premises at any one time or if more than 10 people are at work at any one time elsewhere than on the ground floor. A fire certificate will also be required if explosive or

highly flammable materials are stored or used either in or under the premises.

12 An application for a fire certificate should be made by the owner/occupier of the premises on form FP1(Rev) which can be obtained from the local fire brigade. It is an offence not to apply for a fire certificate if one is required.

13 While awaiting the outcome of the application for a fire certificate the occupier of the premises has a statutory duty to ensure that:

(a) the means of escape in case of fire can be safely and effectively used at all times when there are people in the premises;

(b) the fire-fighting equipment is maintained in efficient working order; and

(c) all staff are trained in what to do in the event of fire.

14 The fire certificate will be tailored specifically to the premises in question. It may impose requirements for maintenance of the means of escape and associated fire precautions, the training of staff, limiting the number of people who may be in the premises at any one time and other precautions that have to be observed.

15 Normally the occupier of the premises is responsible for ensuring adherence to all the requirements of a fire certificate.

16 The law requires that the fire certificate is kept on the premises to which it relates. It should be available for reference or inspection at all times.

17 The fire authority for the area should be informed in advance if it is proposed to:

(a) make a material extension of, or material structural alteration to, the premises; or

(b) make a material alteration in the internal arrangement of the premises or in the furniture or equipment with which the premises are provided; or

(c) begin to store or use, or materially increase the storage or use of, explosive or highly flammable materials.

Premises not requiring a fire certificate

18 Firms that do not employ sufficient people to require a fire certificate and do not use or store explosive or highly flammable materials are subject to section 9A of the Fire Precautions Act 1971 which states that:

> 'all premises to which the section applies shall be provided with such means of escape in case of fire and such means for fighting fire as may reasonably be required in the circumstances of the case.'

Normally the occupier of the premises is responsible for ensuring that this statutory duty is complied with.

Multi-occupied premises

19 In the case of multi-occupied premises the owner will be responsible for the application for a fire certificate and for the provision and maintenance of general fire safety measures. The occupier will, however, normally be responsible for specific fire safety measures, eg keeping exit doors and exit routes available and free from obstruction in the relevant premises and in those common or shared parts of the building over which he has control.

Further guidance

20 Attention is drawn to the publications issued by the Home Office and Scottish Home and Health Department (Refs 6, 7 and 8).

Health and Safety at Work etc Act 1974

21 The HSW Act is concerned with securing the health, safety and welfare of people at work, and with protecting people who are not at work from risks to their health and safety arising from work activities. For further details see References 5 and 9. In terms of fire precautions the HSW Act and relevant Regulations, eg the Highly Flammable Liquids and Liquefied Petroleum Gases Regulations 1972 and the Electricity at Work Regulations 1989, are used to control the keeping and use of flammable substances and to control sources of ignition.

22 Section 6 of the HSW Act applies to

manufacturers, importers and suppliers of substances who have to ensure, so far as is reasonably practicable, that the substances are safe and without risks to health. They need among other things to identify the potential hazards and to provide information about these hazards and the conditions necessary for safe use, for example precautions in handling and storage. In the case of flammable substances this information should include a description of the inherent hazards of fire and the expected consequences if the substance is exposed to fire or excessive heat. The section places similar obligations on designers, manufacturers, importers and suppliers of articles for use at work and on those who erect and install them. Further guidance is available in References 10-14.

23 Inspectors enforcing the HSW Act have the power of entry into premises without prior notice and can issue notices requiring specified improvements or prohibiting the use of substances, plant, equipment or buildings.

24 The enforcing authority should be notified immediately, for example by telephone, if someone is killed or suffers a major injury as a result of an accident arising out of or in connection with work. Immediate notification is also needed where an explosion or fire results in the stoppage of plant or suspension of normal work for more than 24 hours where the explosion or fire was due to the ignition of process materials, their by-products (including waste) or finished products. This has to be followed up by a written report on form F2508. A written report should also be sent following an accident at work if the injured person is incapacitated for normal work for more than 3 days. For details of the Reporting of Injuries, Diseases and Dangerous Occurrences Regulations 1985, see Reference 15.

Petroleum (Consolidation) Act 1928 and Petroleum (Mixtures) Order 1929

25 With certain minor exceptions the Act prohibits the keeping of petroleum spirit unless a petroleum spirit licence is in force (see paragraph 8).

26 Petroleum spirit kept under licence needs to be kept in accordance with the conditions of the licence. A notice setting out any conditions to be observed by employees should be kept prominently posted.

27 Any petroleum spirit kept in any place should have attached to or displayed near the vessel containing it a label showing in conspicuous letters the words 'PETROLEUM SPIRIT - HIGHLY FLAMMABLE'.

28 The Petroleum (Mixtures) Order 1929 applies all of the provisions of the Act, including licensing, to a large range of mixtures whether liquid, viscous or solid. The required label is 'PETROLEUM MIXTURES GIVING OFF A FLAMMABLE HEAVY VAPOUR'.

Building Regulations

29 In England and Wales the Building Regulations 1991 apply to new buildings and to building work such as the erection, extension or material alteration of an existing building. They also apply where there is a material change of use.

30 The Regulations impose fire safety requirements covering such matters as:

(a) means of escape in case of fire;

(b) structural stability;

(c) fire resistance of elements of structure;

(d) compartmentation to inhibit fire spread;

(e) reduction of spread of flame over surfaces of walls and ceilings;

(f) space separation between buildings to reduce the risk of fire spread from one building to another;

(g) access for fire appliances and assistance to the fire brigade.

The standard of provision is related to the size and height of the building and the use to which it is put. In certain buildings the Regulations also impose requirements as to means of escape in case of fire.

31 Where it is proposed to erect a new building, to carry out building work as described in paragraph 29 or to make a material change of use,

application should be made to the building control department of the relevant local authority.

32 In Scotland the Building Standards (Scotland) Regulations 1991 apply. These contain detailed requirements relating to means of escape in case of fire, structural fire precautions and assistance to the fire brigade and, like the Regulations in England and Wales, are related to the size and height of the building and the use to which it is put. The Regulations also contain different requirements for the storage of materials that give rise to different fire hazards.

Dangerous Substances (Notification and Marking of Sites) Regulations 1990

33 These Regulations require those in charge of a site containing 25 tonnes or more of dangerous substances (as defined in the Classification, Packaging and Labelling of Dangerous Substances Regulations 1984) to notify the local fire/enforcing authority and also to erect (specified) warning signs (Figure 1a) at such access points as will give adequate warning to firefighters that dangerous substances are present on a site. Further, from 1 March 1993 site operators will need to erect appropriate warning signs (Figure 1b) at locations within the site where directed to do so by an inspector. For further information see Reference 16.

Classification, Packaging and Labelling of Dangerous Substances Regulations 1984

34 These Regulations provide for the classification, packaging and labelling of dangerous substances for both supply and conveyance by road. They apply, with some exceptions, to substances (as defined in the HSW Act) supplied or consigned for conveyance by road in a package and which are dangerous for supply or for conveyance. See Reference 17 for further details.

Other Regulations and future legislation

35 See References 18-20 for details of other Regulations which may apply to certain firms. European Community Directives 89/391/EEC and 89/654/EEC, known respectively as the Framework Directive and the Workplace Directive,

Figure 1a Access warning sign

Figure 1b Examples of location warning signs

introduce measures to encourage improvements in the health and safety of workers. Both the Health and Safety Commission (HSC) and the Home Departments intend that these Directives be implemented by means of Regulations. In the case of HSC this will be by means of the Management of Health and Safety at Work Regulations 1992 and the Workplace (Health, Safety and Welfare) Regulations 1992 intended to come into force by 1 January 1993. In the case of the Home Departments this will be by Regulations under section 12 of the Fire Precautions Act 1971 covering fire safety measures in existing buildings and coming into force in 1993 with phasing of some requirements. The fire safety measures will be related to a fire risk assessment carried out by the employer. The Regulations will be supported by practical guidance/approved codes of practice.

CAUSES OF FIRES

36 Sources of ignition cannot be completely eliminated. Combustible materials and substances should therefore be controlled both in storage and use. The main causes of fire include:

(a) misused, unsuitable or faulty electrical equipment, including overloads and short circuits not safely cleared by protective devices;

(b) smoking;

(c) poor handling of flammable liquids;

(d) electrostatic sparks;

(e) heating and drying equipment;

(f) frictional heating, eg from hot bearings;

(g) frictional sparks, eg from the use of tools;

(h) poorly maintained equipment;

(i) welding and cutting;

(j) arson.

37 A fire once started will spread due, for example, to:

(a) inappropriate initial action;

(b) poor housekeeping and accumulation of waste material;

(c) unsegregated storage of flammable liquids, solids or gases;

(d) excessive stocks of paper in production areas;

(e) lack of fire separation between floors or rooms;

(f) fire doors wedged or propped open;

(g) inadequate/inappropriate fire detection and extinguishing equipment;

(h) combustible wall and ceiling linings.

FIRE RISK MANAGEMENT

38 The risk of a fire breaking out in a particular place and spreading rapidly will depend largely upon the materials being used and stored, the general standards of housekeeping, the construction and layout of the factory and the training of employees. The risk to people will largely depend on the adequacy and maintenance of means of escape and of the alarm system as well as the training of the workforce in fire routine and evacuation procedures. It follows therefore that for fire risks to be adequately controlled:

(a) the risk of fire occurring is reduced to the absolute minimum;

(b) the risk of fire spreading is minimised; and

(c) everyone is able by their own unaided efforts to reach a place of safety beyond the building (see paragraph 39 for the needs of disabled workers).

39 Employers need to organise for fire safety and should designate a senior member (or members) of the management team to have specific responsibility for fire risk management and staff training. Successful fire risk management entails constant and careful thought, an awareness of all the potential risks associated with the

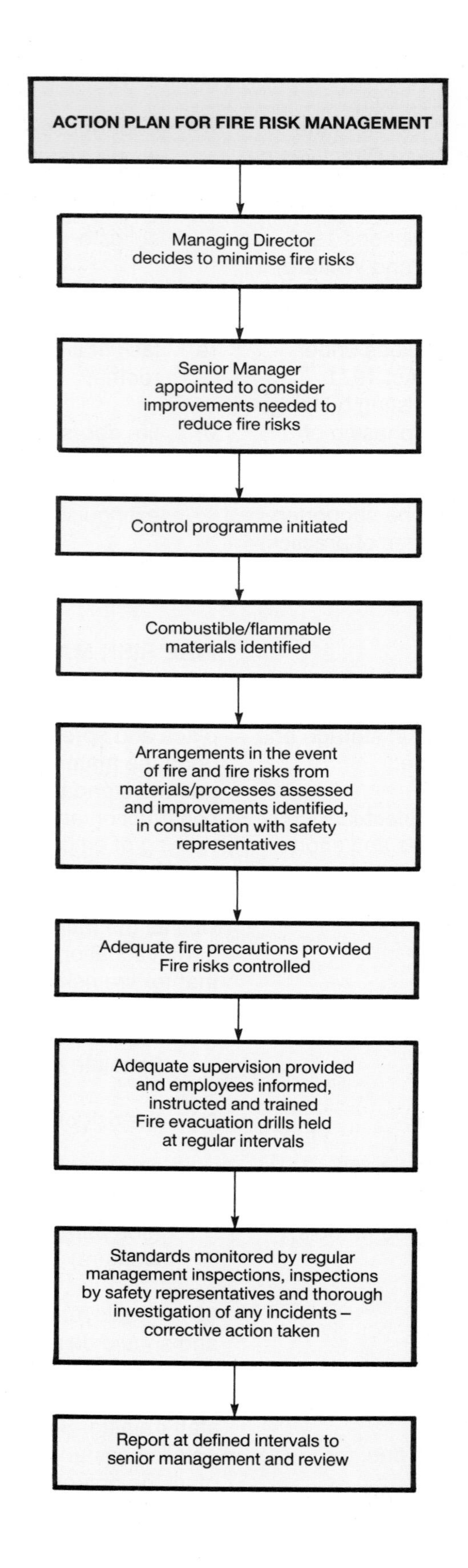

Figure 2 Action plan

premises, its processes, the workforce and contractors, and liaison with the fire authority and insurers. It involves an initial assessment of the risk of fire and the measures provided or needed to combat it, regular periodic surveys of the premises, thorough investigation of any incidents, meetings with key staff and consultation with safety representatives. The special needs of employees who have disabilities and/or sensory impairments should receive particular attention (see References 6 and 21). The arrangements for controlling fire risk should form part of the overall company health and safety policy; for details see References 22 and 23.

40 An action plan for reducing the risk from fire will need to include the following elements as summarised in Figure 2.

Identification, assessment and control

41 Carry out a thorough fire survey in order to assess the fire risks from materials, from the various processes undertaken, and the arrangements in the event of fire breaking out; use of a plan (Figure 3) will be helpful. The survey may be undertaken on a process-by-process basis, eg materials delivery, materials storage, process A, intermediate storage, process B etc, or area by area, eg basement (room A, B), ground floor (room A, B) etc. Ensure that the required improvements are identified, that adequate general fire precautions are provided and that the risks from the processes are properly controlled.

- identify the existing means of escape and ensure these can be used safely;
- find out whether a fire certificate already exists; if not, ask the fire authority whether a fire certificate is required and apply for one if necessary. If a fire certificate already exists, check that it is up to date and all conditions are met;
- check the arrangements for giving warning in case of fire (a fire alarm is a necessity in any building requiring a fire certificate);
- examine the process layout and flow of materials;
- identify materials that will burn readily. Make use of experience of previous incidents and the information which should be available from manufacturers and suppliers (see paragraph 22);
- identify particularly hazardous activities, eg those involving highly flammable liquids, polyurethane foam, welding and cutting, storage of paper reels;
- consider less flammable solvents/water-based inks etc, bearing in mind any health hazards (Refs 1-3, 24 and 25);
- decide whether risks can be reduced by installing safer plant or by modifying existing plant;
- review the arrangements for handling flammable liquids and controlling flammable vapours;
- identify and control sources of ignition including any transient sources from cleaning, maintenance or repair work;
- consider the amount of material both in storage and in use and where it is stored and used - minimise the quantity of flammable material on the premises (set limits) and control its use;
- install fixed fire-fighting equipment where appropriate;
- provide sufficient portable fire extinguishers and/or hose reels;
- identify conditions of fire spread, eg from storage to process areas and vice versa, and separate where necessary;
- consider standards of housekeeping and waste removal, and improve as necessary;
- provide an adequate standard of supervision;
- consider the emergencies that might arise (including spills and leaks) and draw up appropriate procedures;
- review the standard of training given on process risks, precautions and emergency procedures - provide additional training where required;

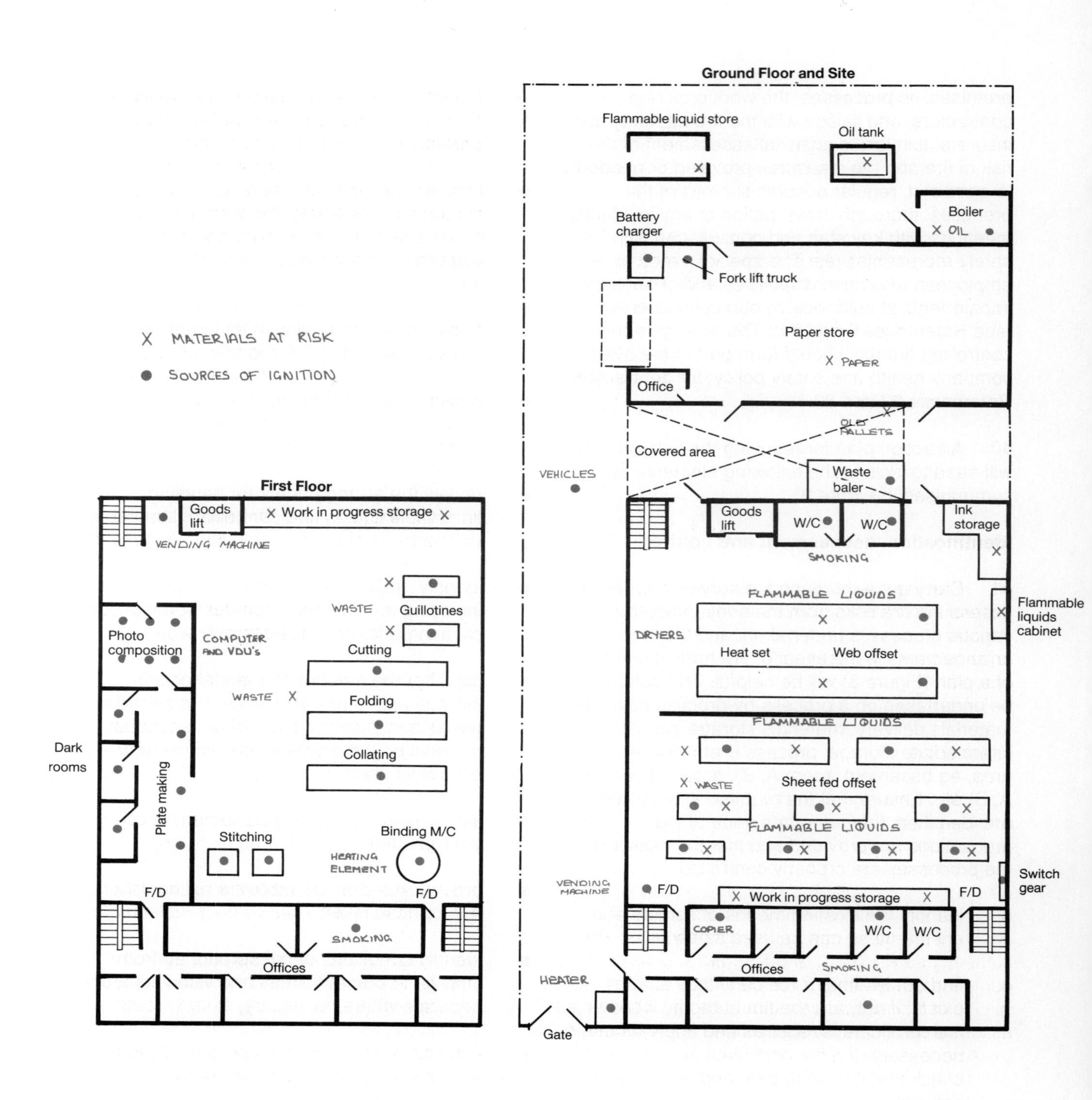

Figure 3 Example of plan prepared during fire assessment

- control smoking;
- review the arrangements for preventive maintenance, eg lubrication, electrical checks etc;
- consider the activities of contractors and provide adequate controls, for example by preparing and enforcing rules for contractors doing hot work and on minimum standards for electrical equipment brought onto site;
- assess the adequacy of fences, doors, etc in keeping out children and other intruders;
- obtain expert fire advice prior to any material alteration to the premises or processes carried on.

Training

42 Provide adequate supervision and inform, instruct and train employees in the process risks, the standards to be maintained, working practices to be adopted, fire routine and evacuation procedures (see paragraphs 44 and 66 and Reference 26).

Monitoring

43 Maintain standards by monitoring to see what is being achieved in practice and correcting any shortcomings (Ref 27). While all employees should report any hazard they notice, a system of regular formal inspection by management is necessary to monitor conditions and to identify changes that may introduce additional risks. The company safety policy should make it clear who carries out these inspections and when. The results of the inspections should be recorded to aid assessment and reports at defined intervals made to senior management for review.

GENERAL FIRE PRECAUTIONS

Training

44 In order to minimise the risk to people in case of fire, it is essential that they all receive adequate fire safety training appropriate to their role. Fire safety training can be broadly divided into four types:

(a) *Induction training* should be given to all new staff on their first day and should include an explanation of evacuation procedures, method of raising the alarm and any rules concerning smoking. There should be a tour of escape routes from the premises during which time fire alarm call points, fire equipment and fire doors should be pointed out.

(b) *Basic training* should be given to all staff, preferably at least twice, and in all cases not less than once a year. The training should provide for the following:

  (i) the action to be taken on discovering a fire;

  (ii) the action to take on hearing the fire alarm including evacuation and roll-call procedures;

  (iii) how to raise the alarm and the location of fire alarm call points;

  (iv) how the fire brigade is called;

  (v) the location and use of fire equipment and the dangers of using the wrong type of extinguisher;

  (vi) knowledge of escape routes including any stairway not in regular use;

  (vii) knowledge of the method of operation of special emergency door fastenings;

  (viii) location and identification of fire doors and their importance in restricting fire spread and protecting escape routes;

  (ix) stopping machines and processes and isolating power supplies where appropriate;

  (x) warning against stopping to collect belongings or re-entering buildings.

(c) *Training of key personnel* should apply to certain categories of staff who will need to be instructed and trained in matters which will be their particular responsibilities over and above basic training. Such staff will include:

(i) heads of departments;

(ii) engineering and maintenance staff;

(iii) supervisors;

(iv) security staff;

(v) wardens/marshalls (where appropriate);

(vi) safety representatives;

(vii) telephonists.

Specific aspects of training will include the supervision of evacuation and roll-call procedures, the control of contractors and the safety of members of the public in the event of fire. Managers of large establishments may consider giving additional training to a small number of staff to enable them to fight a fire safely and efficiently as a team in an emergency until the fire brigade arrives; if this is proposed they should first consult the fire authority. See also Reference 28.

(d) *Refresher training* should be given to appropriate staff at such intervals as to ensure that changes in procedures, any new hazards or any lessons learned from fire incidents or drills are made known to those concerned.

Note: See paragraph 66 for training in the specific precautions associated with the processes undertaken.

45 A practice fire drill should be carried out at least once and preferably twice a year simulating conditions in which one or more exits or escape routes from the building are obstructed. During these drills the fire alarm should be operated by a member of staff who is told of a supposed outbreak of fire. The fire routine should then be followed as fully as circumstances permit.

46 All training and drills should be based on written instruction and be recorded in a fire log book. The records should include:

(a) the date of the instruction or drill;

(b) duration;

(c) name of the person giving the instruction;

(d) names of the people receiving the instruction;

(e) the nature of the instruction or drill;

(f) any observations/remedial action.

47 Printed fire instruction notices should be displayed at conspicuous positions in the building stating in concise terms what staff and others should do if a fire is discovered or if they hear the alarm. The notices should be permanently fixed in position and suitably protected to prevent loss or defacement. If a fire certificate is required for the premises, it will normally specify the content of the notice.

Means of escape in case of fire

48 All premises should have means of escape which are appropriate to the risk. The means of escape provided should be maintained unobstructed and should be available for use for the whole of the time that the premises are occupied. Doors should be capable of being easily and immediately opened from the inside.

49 Circumstances in each premises will vary, but the principle to be applied is that a person should be able to turn away from a fire and reach a place of safety, in the open air, within a reasonable distance and without outside assistance. Arrangements for disabled people should be carefully considered (see paragraph 39).

50 To be taken into consideration when assessing means of escape are such matters as:

(a) fire resistance and surfaces of walls and ceilings;

(b) assessment of fire risk;

(c) distance of travel;

(d) number of floors;

(e) number of stairways;

(f) arrangement of machinery, fixtures, fittings etc;

(g) lighting of escape routes;

(h) signs and notices.

51 Most printing industry premises will require a fire certificate. If a fire certificate already exists, the means of escape detailed on the certificate should be properly maintained.

52 Pending the issue of a fire certificate the existing means of escape should be maintained so that they can be used safely, ie free from obstruction, doors capable of being easily and immediately opened from the inside. Premises not required to have a fire certificate should have adequate means of escape. For further information see References 4 and 6-8.

Fire alarms

53 A fire alarm system is always to be recommended, but in any building or part of a building requiring a fire certificate a fire alarm is a legal requirement.

54 Other than in very small premises a type M system (manually operated electrical fire warning) as described in British Standard 5839: Part 1 (Ref 29) will generally be considered to be the minimum requirement. Among the matters to be determined are the provision and siting of call points, the siting of alarm sounders and of control and indicating equipment. If alterations to plant or buildings are made, the audibility of the alarm should be checked.

55 Automatic fire detection (eg smoke and/or heat detectors) is desirable. Where manually operated electrical systems and automatic systems are installed in the same building, they should be incorporated into a single integral system (Ref 29).

Automatic sprinklers and automatic smoke control systems

56 Automatic sprinklers, automatic smoke control systems (see paragraph 69), or a combination of both can control fire and restrict fire spread. Such matters will require specialist advice which should be sought from the fire service, insurers or from firms having appropriate expertise.

Gaseous fire extinguishing systems

57 These are commonly used to protect certain printing presses, computer suites and data stores from fire. Exposure to the extinguishing agent is potentially hazardous in the event of accidental discharge, as is exposure to the agent and products of combustion during discharge in a fire.

58 These systems should be installed in accordance with the guidelines contained in Reference 30. Generally they should be under manual control when people are present. Warning signs should be provided and in the case of a total flooding system there should be an indicator showing whether the system is on manual or automatic. Adequate means of escape should be provided and maintained. Further guidance is available in BS 5306: Parts 4 and 5 (Ref 31).

Extinguishers and hose reels

59 All premises should be provided with means for fighting fire for use by people in the building; see BS 5423 *Specification for portable fire extinguishers* (Ref 32).

60 In selecting appropriate means for fighting fire the nature of the materials likely to be found should be considered. Different types of fires are defined in BS 4547 *Classification of fires* (Ref 33). These and the appropriate extinguishing agents are described in Appendix 2.

61 People on the premises should be aware of the dangers and limitations of fighting a fire with fire extinguishers.

Maintenance and testing

62 Appropriate arrangements should be made for maintaining and testing the fire alarm system (normally a three-monthly test), emergency lighting and fire-fighting equipment. A record should be kept in the fire log book (see paragraph 46). The operation of all escape doors not in regular use should also be checked periodically.

Note: The general fire precautions outlined in paragraphs 44-62 are covered by the Fire Precautions Act 1971 and are normally enforced by the fire authority (see paragraph 6).

HOUSEKEEPING

63 A programme of regular cleaning and waste removal needs to be set up. The frequency will vary between companies but it should be done at least once per day. Cleaning must be thorough and include office and storage areas. Particular attention should be paid to those areas where paper dust or fine trim is produced as this is easily ignited. Normally a vacuum cleaner should be used, not an air line or a brush and shovel.

64 A sufficient number of waste containers, preferably non-combustible, should be provided at appropriate points. There should be a system for the prompt removal of the large volumes of waste liable to be generated at such machines as web-fed presses, inserting machines and slitting/trimming machines. Waste extraction equipment should be employed where possible.

HAZARDS AND PRECAUTIONS IN VARIOUS OPERATIONS

65 Management should look at the entire factory, warehouse and office operation and identify the hazards in the various departments. The precautions and standards outlined will help management to monitor the company's effectiveness in combating these hazards.

66 In addition to the training in general fire precautions outlined in paragraph 44 all those on site should be informed of the hazards from the flammable material stored there and of the need to isolate sources of ignition and heat. Those working in process and storage areas should also receive specific training in both normal operating procedures and emergencies. Periodic retraining will also be necessary. Training should, where appropriate, cover:

(a) safe operation of plant and equipment, safe handling of flammable materials (see paragraph 87 for organisations providing training in the hazards of flammable liquids) and safe systems of work for cleaning and maintenance;

(b) housekeeping;

(c) reporting of faults and incidents, including minor leaks and spills of flammable liquid;

(d) emergency procedures in the event of fire, spills and leaks;

(e) relevant legal requirements.

Welding and cutting

67 Welding and cutting etc should be under a written 'hot work permit', which includes an initial assessment of the area and clearing the immediate vicinity. See References 34-36 and Appendix 3.

Storage

68 The main principles of safe storage are to segregate the storage areas from the process work areas and segregate flammable materials so that in the event of a fire in one area, other areas are not affected. Only authorised people should be allowed access to fire-separated storage areas.

69 One method of reducing the risk of fire and smoke spread and of controlling the fire size is to ensure that it is vented in its early stages. This may be achieved by automatic smoke venting and smoke curtains where necessary or the provision of a roof which will fail early in any fire. Such a provision has the added advantage of allowing firefighters to enter the building and locate the seat of the fire.

70 While storage in a separate building, a single-storey extension to the main building or in a safe place in the open air is to be preferred, this may not always be possible. In these cases storage areas should be separated from workrooms by partitioning having a fire resistance of not less than half an hour. If the storage is in the same building as residential accommodation, there should be fire-resisting separation of not less than one hour between them.

71 In certain other limited cases where storage areas cannot be separated, safety will need to be achieved by other means (but see paragraph 72). This will require a careful assessment, possibly drawing on expert advice, of the characteristics and quantity of material stored, conditions of use and storage, the nature of the building and the adequacy of emergency procedures and fire precautions. For example, it may be possible to achieve safety by a combination of other measures such as spatial separation, fire or smoke detectors

linked to an alarm system, sprinklers and smoke control systems. Although such arrangements may be reasonable to protect employees, it must be remembered that lack of physical separation can lead to more extensive fire damage.

72 Separation may be specified in a fire certificate or by the conditions of a petroleum licence. Separation may also be required where the Highly Flammable Liquids and Liquefied Petroleum Gases Regulations 1972 apply.

73 The following points are of particular note:

(a) a separate fire-resisting storeroom should be provided for foam plastics and other materials that may give off particularly toxic fumes in a fire (Ref 37);

(b) gangways in stacking areas should be sufficiently wide to ensure free movement to fire exits; a currently accepted minimum figure is 1.1 metres. Where mechanically powered or hand trucks are used in the gangway, it should be of sufficiently greater width to leave enough space for people to pass freely. Gangways should also be clearly marked, for example by lines painted on the floor, and kept clear. An exit should be provided from each end of any main gangway and dead ends should be avoided;

(c) stored combustible material should be kept well away from potential sources of ignition such as light fittings and direct-fired or convector-type heaters. Raw materials and products should not be stored so as to obstruct means of escape, fire alarm call points, fire detection equipment, fire-fighting equipment (including sprinkler heads) or fire doors and shutters;

(d) smoking should be prohibited in the store;

(e) boilers, emergency generators and other similar plant should be segregated by fire-resisting walls, floors and doors;

(f) lift truck battery charging should be excluded where possible. If this cannot be achieved, a separate dedicated bay should be provided or charging operations separated from stored material by a safe distance (preferably at least 3 metres). See also paragraph 95;

(g) used pallets and empty containers stored outside should be well away from openings into buildings and preferably at least 1 metre clear of any boundary fence;

(h) solvent impregnated paper should be kept in a fire-resisting store. Solvent impregnated waste paper and rags should be placed in non-combustible containers with close-fitting lids and disposed of regularly;

(i) vigilance against arson is needed, especially where there are very large quantities of stock and few employees. Specialist advice on security matters may be needed;

(j) further guidance on controlling fire hazards arising from storage activities can be found in References 14 and 38.

Reel storage

74 Storage of paper reels can create particular fire hazards and it is especially important that the principles outlined in paragraphs 68-73 and in Reference 14 are followed. Management should assess the risk arising from such storage and see that the necessary safeguards and controls are applied, taking account of information provided by the suppliers (see paragraph 22). See also paragraph 43 and Reference 27.

75 The most likely material to be ignited is loose paper material, eg discarded wrappers, loose ends, and loose paper on damaged reels. These problems need to be identified at an early stage. Damaged reels should be wrapped or taped on delivery.

76 Where reels are stacked vertically in columns there will inevitably be voids between the columns. In the event of a fire starting at the bottom of the stack the gap will constitute a chimney. Rapid burning will follow as air drawn in at the bottom will accelerate up the chimney, typically to a speed of approximately 15 metres per second. The rate of burning increases rapidly with increasing stack height. This rapid vertical flame spread is compounded by rapid horizontal spread when the layers of paper begin to unwind. Special care should be taken when considering the installation of sprinklers if these systems are to be able to contain such fires (see also British Standard 5306: Part 2 (Ref 31) and References 39 and 40). The

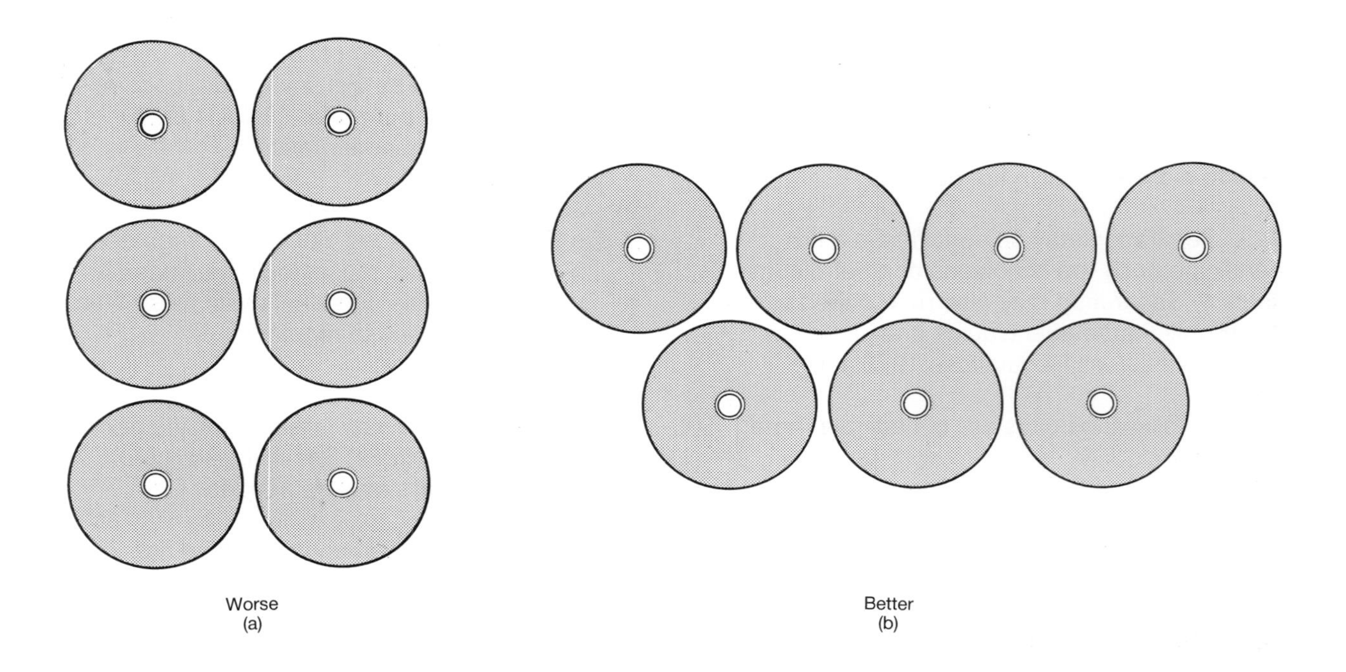

Figure 4 Patterns of vertical reel stacking viewed from above

risk of rapid fire spread is reduced if reels are not stacked on end. Vertically stacked paper reels should not normally be stored below an occupied floor or in a basement.

77 The following special precautions are required in reel storage areas:

(a) floors need to be kept clean and free of loose paper. Other combustible material should not be kept in the reel store;

(b) damaged reels should be repaired by taping or rewrapping;

(c) when reels are stacked vertically the stack height should be minimised and the spaces between stacks should be as small as possible. Spaces of less than 100 mm should help prevent the peeling off process that accelerates fire spread. If spaces can be reduced to less than 25 mm, the airflow through the stack is likely to be throttled and the rate of fire spread further reduced; where this close stacking is being considered it is essential that the layout and method of stacking and retrieval will allow for this reduction of space without risk of dislodging reels in adjacent stacks. Alternatively the spaces should be wide enough to allow access, ie normally at least 1.1 metres. It is particularly important to restrict pedestrian access where reels are stacked vertically because of the hazards of fire spread and of dislodging reels. Figure 4 shows two stacking arrangements in which the distances between reels are equal. The arrangement in Figure 4b is preferable; the voids between reels are smaller and less well ventilated;

(d) smoke venting etc should be provided (see paragraph 69);

(e) sources of ignition should be rigorously controlled.

78 Stacked reels can present an unacceptable risk to people in inner rooms such as offices, storerooms, darkrooms, toilets, etc. The presence of stacked reels in the vicinity of the only escape route from these rooms raises the outer room (ie store) to high risk status. Unless reels are sufficiently remote from the escape route from any inner room, then the inner room should have an alternative exit either direct to open air or via a separate fire-resisting compartment. The general principles of means of escape from inner rooms described in paragraph 98 apply. For further

information see the Home Departments' guide (Ref 6).

79 The stacking of reels can have an adverse effect on the audibility of the fire alarm system. If stacking is introduced into an area not previously used for that purpose or if there is any material increase in such stacking, a check should be made to ensure that the alarm is still audible throughout the whole premises.

Gases

80 Gases in cylinders can be flammable, toxic, corrosive or inert. Even cylinders of inert gases pose a hazard in a fire as they can become over pressurised and rupture violently.

81 All gas cylinders should be stored away from flammable liquids, combustible materials, corrosives and toxic materials. Cylinders containing gases of differing hazard, eg toxic, flammable, corrosive and oxidising (including oxygen), should be stored separately from one another. The main points of note are:

(a) gas cylinders should preferably be stored in the open air in a lockable wire mesh cage for security. For details of separation distances from buildings and boundaries see the references given in paragraph 82;

(b) except for small quantities, cylinders of liquefied petroleum gas, eg propane and butane, should be kept separately from other gases;

(c) if flammable gases are to be used in a building, the preferred arrangement is for the cylinders to be kept in the open air and piped into the workroom by permanent fixed metal pipework at as low a pressure as is possible.

82 Detailed guidance is available in References 41-46.

Flammable liquids

General

83 These will include many blanket washes, revivers, ink removers, alcohols for damping systems, coatings and gravure and flexographic inks containing solvents, as well as paraffin and white spirit. The Highly Flammable Liquids and Liquefied Petroleum Gases Regulations 1972 apply to those liquids having a flash point of less than 32°C and which support combustion, ie at ambient temperature they give off vapours that are easily ignited by, for example a small spark or hot surfaces; in some cases the Petroleum (Consolidation) Act and the Petroleum (Mixtures) Order will apply. The flash point should be given in the supplier's health and safety data sheet.

84 The main hazard in the storage of flammable liquids is fire involving either the bulk liquid or escaping liquid or vapour; the vapours are heavier than air and can travel long distances so that any major spillage will almost inevitably reach a source of ignition. Such incidents may be caused by inadequacies in design, manufacture, installation, maintenance, or by equipment failure or maloperation, together with exposure to a nearby source of ignition. Ignition of the vapours from flammable liquids remains a possibility until the vapour concentration is reduced below the lower flammable limit (LFL). The LFL of a material is the concentration (usually expressed as the percentage, by volume, of the material mixed with air) below which the mixture is too lean to undergo combustion; this varies with different materials but it is usually about 1.5% in air.

85 Examples of hazards in the use of flammable liquids are fire following the ignition of vapour from a leak or spill or escaping from a poorly enclosed process, or explosion from the ignition of vapour within an extraction system, dryer or enclosed space. Such incidents can be caused through handling in open containers, inadequate conveyance or processing arrangements or those shortcomings identified in paragraph 84; serious accidents resulting from initiation ceremonies or horseplay have also occurred.

86 The safe use of flammable liquids is discussed in later paragraphs dealing with specific processes; guidance is also available in Reference 47. Precautions for dryers are set out in paragraph 126 (see also References 48 and 49 and Appendix 6). The following general points apply wherever flammable liquids are used or kept:

(a) stocks of flammable liquids and empty or part-used containers should be properly stored (see paragraphs 89-92);

(b) the quantity of flammable liquids in workrooms should be kept to a minimum, normally to no more than a half-day's supply;

(c) flammable liquids should be conveyed and handled in enclosed systems where possible, for example by piping supplies from the storage location to the point of use; where a connection is frequently broken and remade, a sealed end coupling is preferred. Containers should be kept covered. Special purpose safety containers with self-closing lids and caps should be used where possible for dispensing and applying small quantities of flammable liquids (Figure 5). The lids and caps of containers should be replaced after use. Containers should not be opened in such a way (eg by punching a hole or cutting off the top) that they cannot be reclosed;

(d) any dispensing or decanting should be carried out in a safe place. Spill trays or other means to contain spillage should be provided;

(e) rags impregnated with flammable solvents should be kept in metal bins with well-fitting lids (Figure 6) and disposed of promptly;

(f) sources of ignition should be controlled (see paragraph 88);

(g) flammable liquids should be stored and handled in well-ventilated conditions. In some cases it may be necessary to use exhaust ventilation to control flammable vapour.

87 Key personnel should be trained as necessary in the characteristics and dangers of flammable liquids; this could involve attending a short course, for example those run by the Petroleum Training Federation, 162-168 Regent Street, London W1R 5TB and by some fire authorities.

Sources of ignition

88 There should normally be no means of igniting flammable vapour within hazardous areas associated with the use and storage of flammable liquids. Potential sources of ignition include naked flames (see paragraph 67), hot surfaces, unprotected electrical equipment, static electricity and vehicles. The following points are of particular note:

(a) areas where electrical equipment is located and where flammable vapour may be present should be classified into the following hazardous zones in accordance with BS 5345 (Ref 50):

Zone 0 - in which an explosive atmosphere is continuously present, or present for long periods;

Zone 1- in which an explosive atmosphere is likely to occur in normal operation;

Zone 2 - in which an explosive atmosphere is not likely to occur in normal operation, and if it occurs, it will exist only for a short time.

Areas outside these zones are defined as non-hazardous. Electrical equipment should be positioned in non-hazardous areas where possible. Any electrical equipment in a hazardous area should be suitable for use in flammable atmospheres (Ref 51); when determining these areas References 47, 52 and 53 may be consulted. The areas should be reviewed regularly and recorded on a works plan;

(b) smoking should not be permitted where flammable liquids are handled. No smoking signs should be posted;

(c) electrostatic discharges can ignite flammable vapour. Charges can be generated by, for example, the movement of a printing web, the movement of organic solvents, by people not wearing anti-static footwear or by people removing outer garments of synthetic material. See paragraph 124 for further details of charging mechanisms and precautionary measures;

(d) vehicles such as lift trucks operating in areas where a flammable vapour may be present should be suitably protected (see Reference 54 for guidance on the use of diesel-engined lift trucks in hazardous areas);

(e) frictional sparks generated by, for example, the use of hand or power tools can ignite flammable vapour. Flammable liquids residues and vapour should be removed as far as possible before doing work that may cause frictional sparks (even 'reduced

Figure 5 Examples of special purpose containers for flammable liquids

Figure 6 Example of metal container for cloths contaminated with flammable solvents

sparking' tools can cause sparking due to embedded grit or tramp metal);

(f) indirect means, eg hot water radiators, should preferably be used to heat workrooms where flammable liquids are handled. Portable heaters should not normally be used;

(g) fan motors used at ventilation ducting should not be located in the path of the vapour being removed. A centrifugal fan, a motor with belt drive to the fan or bifurcated ducting with a centrally mounted fan may be used. The duct work should be fire-resisting.

Storage

89 The detailed advice on storage of flammable liquids given in References 55-57 should be followed (certain variations for liquids with flash points in the range 32-55°C are given in these publications).

90 The main requirements for bulk storage in fixed tanks are:

(a) bulk tanks should generally not be sited inside a building, nor should they be sited on the roof of a building but in the open air in a secure well-ventilated position. Recommended minimum separation distances to buildings, boundaries, etc are given in Table 1;

(b) tanks of flammable liquids should be surrounded by an impervious bund wall that is large enough to contain the contents plus 10% of the largest tank but is no higher than 1.5 metres. Arrangements should be made for the safe removal of rainwater;

(c) tanks can be sited underground but they should be placed at least 2 metres outside the building line and not below the floors of buildings.

91 Any container larger than about 1000 litres nominal capacity and connected directly to a process or other point of use, for example a framed and skid-mounted container, should be considered to be a fixed tank.

92 The main requirements for the storage of portable containers are:

(a) recommended minimum separation distances for drums and similar portable containers, including empty containers (unless made free of vapour) are given in Table 2. Where adequate separation distances cannot be achieved, a fire wall of at least half-hour fire resistance (Appendix 4) may be used (see Figure 7). The wall may form part of a bund wall, building wall or boundary wall;

(b) where drums and containers are kept inside a storeroom it should, wherever possible, be

Table 1 Minimum separation distance for bulk storage tanks containing highly flammable liquids

Tank capacity		*Minimum separation distance from any part of a tank*	
Single tank	*Total for a group (maximum)*	*From building, boundary, source of ignition, filling point or process unit*	*From a bund wall*
Metres^3	*Metres^3*	*Metres*	*Metres*
Up to 1	3	1*	1
1-5	15	4	1
5-33	100	6	1
33-100	300	8	1
100-250	750	10	2
Above 250	-	15	2

* But at least 2 metres from doors, plain-glazed windows, ventilation or other openings or means of escape. Also not below any opening, or means of escape, from an upper floor regardless of vertical distance

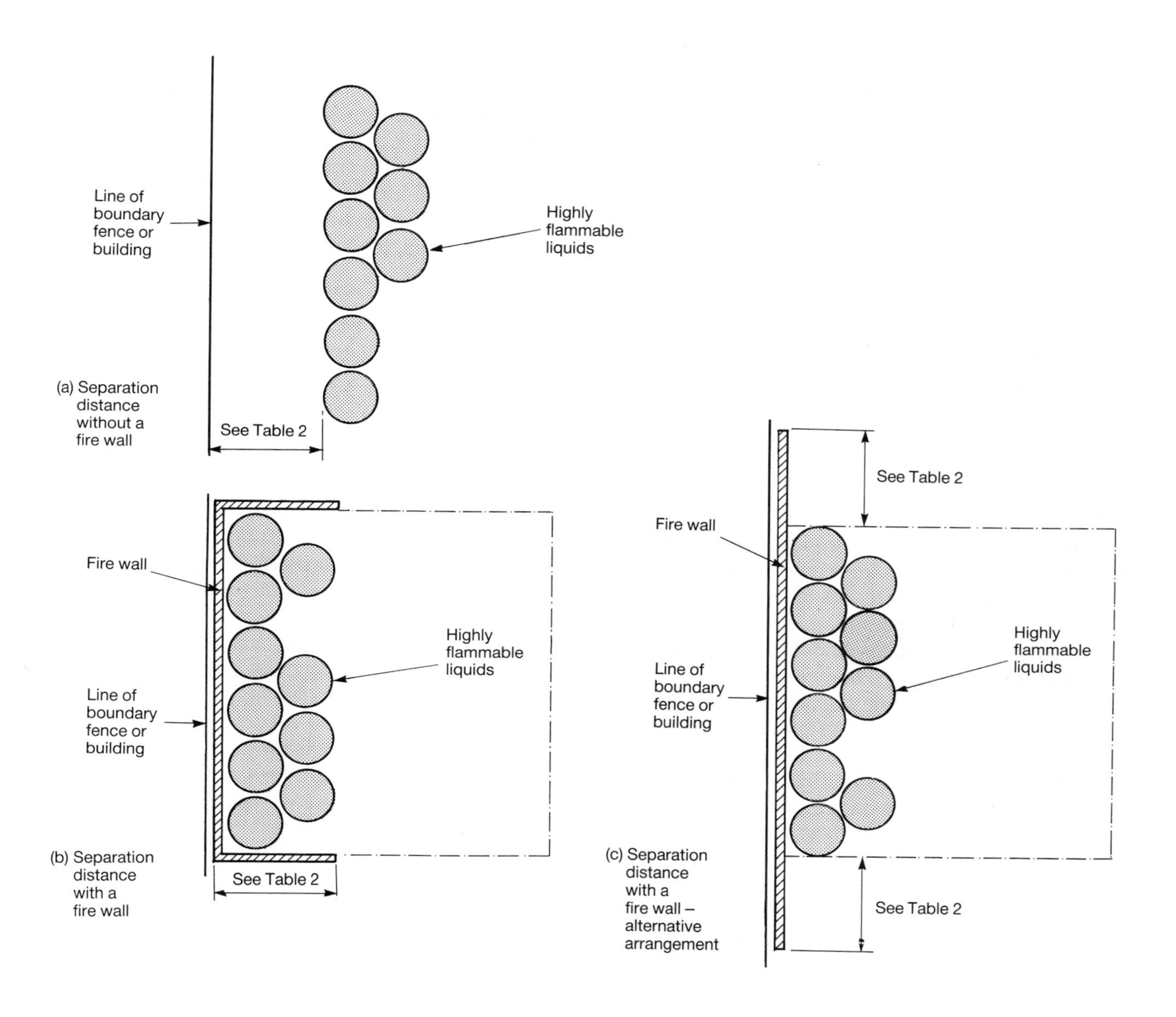

Note: Fire wall to be at least as high as container stack, with a minimum height of 2m, and sited within 3m of stack

Figure 7 Separation distances for highly flammable liquids in drums and similar portable containers stored outside (viewed from above)

Table 2 Minimum separation distance for containers of highly flammable liquids

Quantity stored	*Distance from occupied building, boundary, process unit, flammable liquid tank or fixed ignition source*
Litres	*Metres*
Up to 1000	2
1000 - 100 000	4
Above 100 000	7.5

a secure separate building situated away from adjacent buildings, process units or boundary fences. Where the storeroom is not remote from other buildings or boundaries, it should have a structure giving at least half-hour fire resistance;

(c) storerooms should be adequately ventilated to disperse vapour from any leakage or spillage. A standard equivalent to at least five air changes per hour is recommended using low and high level air bricks to give good natural ventilation. The openings for ventilation should have a total area of at least 1% of the total area of the walls and roof. In a store which forms part of a larger building, the store should preferably have at least two external walls with natural high and low level ventilation provided only in these walls (Figure 8). Ventilation openings should not normally be provided in internal walls. See also Reference 56;

(d) sources of ignition including portable equipment such as welding sets and handlamps should be kept out of any storerooms. Electrical equipment should be suitable for use in the circumstances, ie normally protected to a Zone 2 standard (see paragraph 88);

(e) provision should be made for containing any spillage either by providing a low (typically 150 mm) sill at the entrance, a sump in the storeroom or a retaining tray. The containment should be capable of retaining the contents of the largest vessel plus 10%;

(f) external storage buildings, whether they are fire-resisting structures or not, should have a lightweight roof to act as relief in the event of an explosion;

(g) the store should be used exclusively for the storage of flammable liquids and should not normally be used for dispensing;

(h) up to 50 litres of liquids may be kept within the workroom provided they are in a suitably placed fire-resisting bin or cupboard (Figure 9). See also Appendix 4.

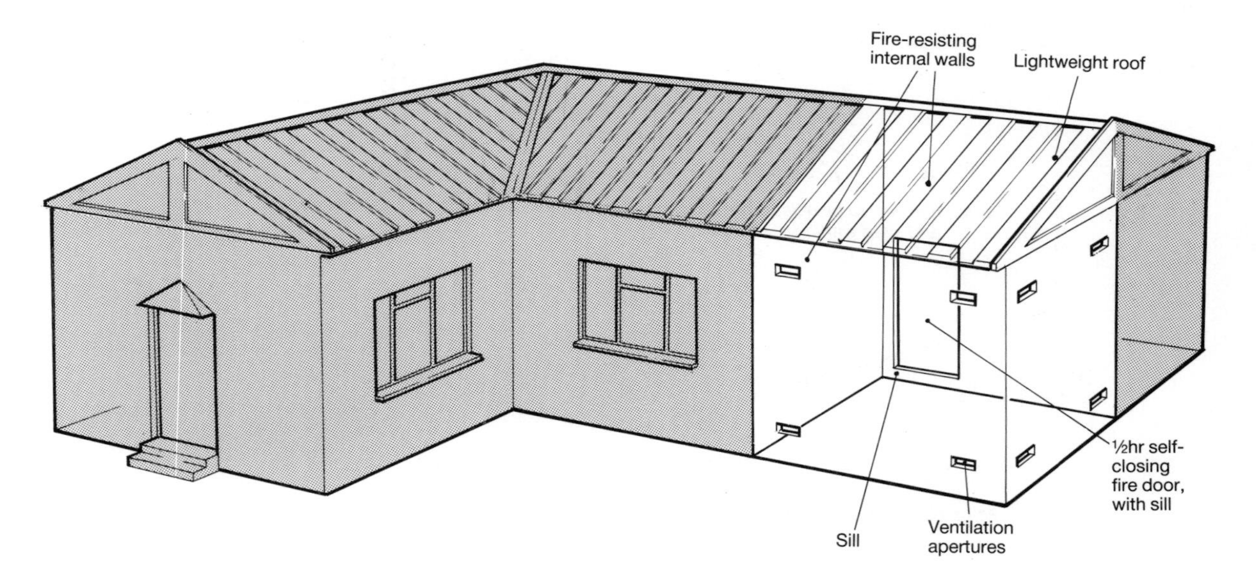

Figure 8 An example of a suitable storeroom in a building. In the case of a multi-storey building advice should be sought from the relevant enforcing authority

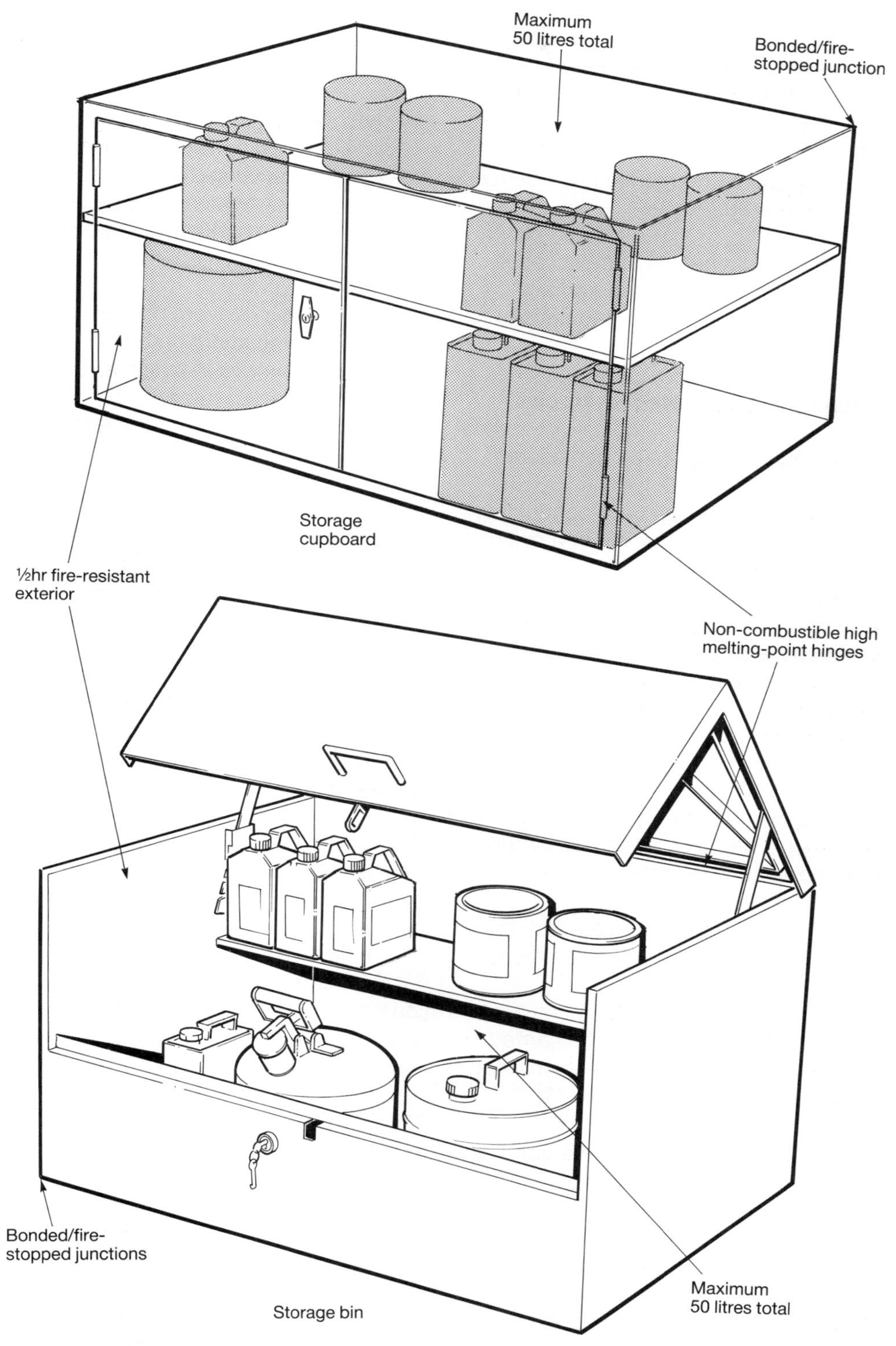

Note: Cupboard or bin to be labelled in accordance with paragraph 93

Figure 9 Storage in the workroom

Labelling

93 Storerooms, cupboards or bins should be clearly marked 'Highly flammable' or 'Flash point below 32°C' (or 'Flammable liquid' in the case of liquids with a flash point in the range 32-55°C). Petroleum stores should be marked in accordance with paragraphs 27 and 28.

94 The labelling of containers is in most cases dealt with by the Classification, Packaging and Labelling of Dangerous Substances Regulations 1984 (see paragraph 34).

Charging of electrically operated lift trucks

95 The following points need to be considered:

(a) if a metal tool or other electrically conducting object short circuits the terminals of a cell or cells, it will become hot and may cause burns. In addition sparks and molten metal may be ejected. Insulated tools should always be used and, before working on a battery, people should remove any metallic items from hands, wrist and neck, together with any such items that may fall from pockets;

(b) hydrogen and oxygen are emitted from a battery when it is being charged. Hydrogen/air mixtures produce violent explosions if ignited and it has to be assumed that this mixture is present in the immediate vicinity of the cell tops at all times. To minimise the risk of explosion:

 (i) charging should preferably be carried out in an area used exclusively for that purpose (see also paragraph 73(f));

 (ii) the charging area should have good natural high level ventilation immediately above the batteries;

 (iii) light fittings should be of a totally enclosed industrial type, eg bulkhead fittings, and preferably positioned other than directly above the charging area;

 (iv) smoking or naked lights should be prohibited in the area. Appropriate notices should be displayed;

 (v) anything capable of causing a spark should not be used in the vicinity of the cell tops;

 (vi) a proper plug and socket arrangement should normally be used for routinely connecting the charger to the batteries - the charger should be switched off before making or breaking the connection;

 (vii) battery covers may be open or removed during charging but all vent plugs should be in position;

 (viii) when carrying out maintenance on the battery, all electrical circuits including the charger should be switched off before making or breaking connections at the battery terminals. The lead connected to the vehicle framework should always be disconnected first and reconnected last;

(c) appropriate fire-fighting equipment should be readily available (see Appendix 2).

Pre-press

96 This includes typesetting, film and plate processing and screen make-up, some of which may involve computerised operations.

97 A likely cause of fire in these areas is unsatisfactory electrical equipment. Obvious sources of heat, such as lamps on cameras, should be kept well clear of combustible material, including partition walls.

98 Particular attention should be paid to the means of escape in case of fire from darkrooms. Before a darkroom is provided or the layout of any darkroom is changed, the fire authority should be consulted on the adequacy of the means of escape. If the maximum travel distance within a darkroom to the darkroom exit is more than 12 metres, generally an alternative exit will be required. Darkrooms preferably should have escape routes via a fire-protected corridor or direct to open air. However, where the darkroom is an inner room (Figure 10), ie a room only accessible through another room:

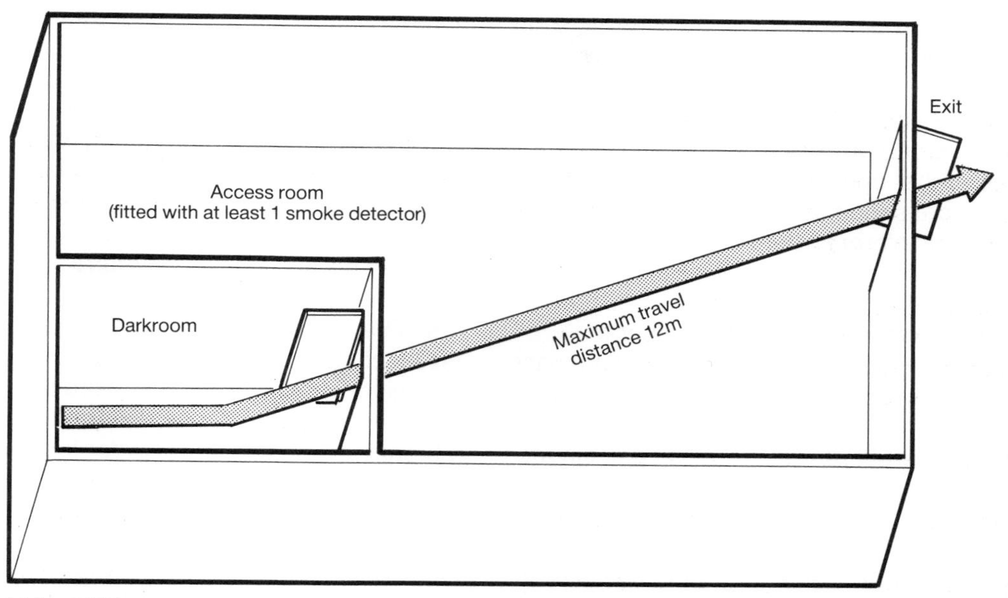

(a) Acceptable

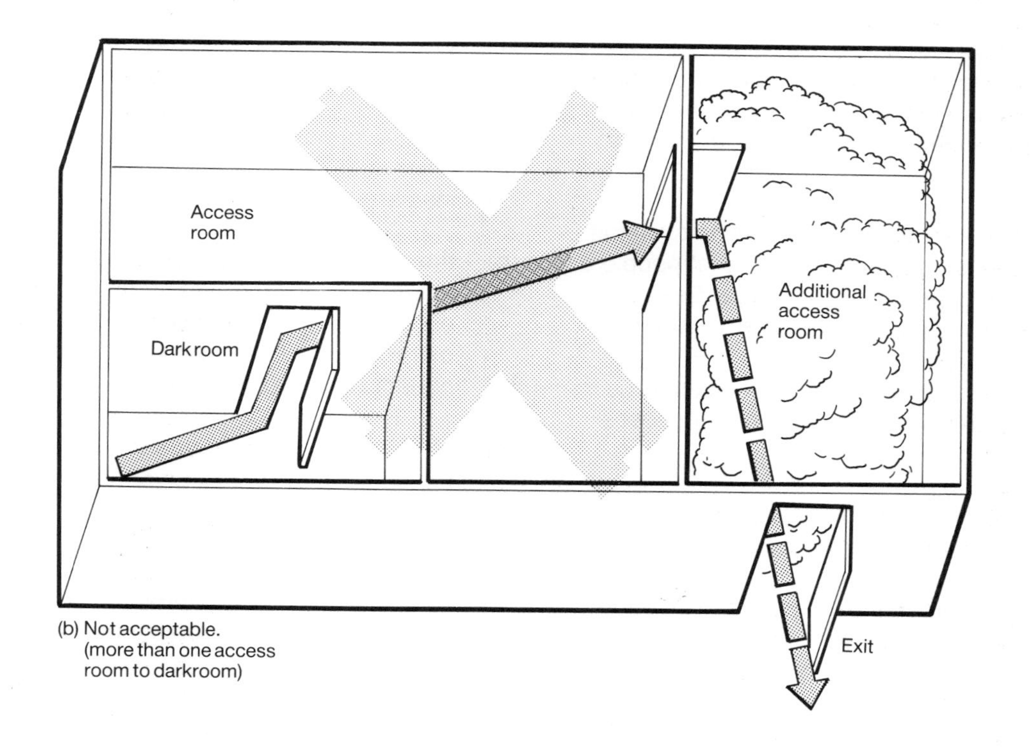

(b) Not acceptable.
(more than one access
room to darkroom)

Figure 10 Means of escape from darkroom via access room

(a) the access room should not be an area of high fire risk;

(b) as clear vision panels cannot in practice be used at darkrooms, at least one smoke detector that operates an alarm audible in the inner room should be fitted in the access room. This alarm system should conform to the requirements of BS 5839: Part 1 (Ref 29);

(c) escape from an inner room to either a protected route or a final exit should not pass through more than one access room; and

(d) the total travel distance from any point in the inner room to the access room exit should generally not exceed 12 metres.

99 Floors and work surfaces should be cleaned regularly and off-cuts of film, paper and other material should be placed in bins, preferably non-combustible, which should be emptied at least daily.

100 Flammable liquids such as film cleaners or adhesives should be correctly stored and used, and only the minimum quantity needed for use kept in the workroom (see paragraph 92).

101 Where gaseous fire extinguishing systems are used, care is needed to ensure that people are not trapped in the event of discharge. See paragraphs 57 and 58 and Reference 30.

102 Solvent recovery systems are used in association with some plate-making processes (see Appendix 5).

103 Grinding or polishing of magnesium printing plates is a particularly hazardous process subject to specific legal controls (Ref 58). A number of fires have occurred. A separate dust extraction system which includes a wet scrubber with explosion relief should always be provided. Grinding or polishing of iron or ferrous metals should not be undertaken on equipment used for magnesium. Smoking should be prohibited and proper arrangements made for disposal of waste. The correct fire-fighting equipment should also be available (see Appendix 2, paragraph 7).

104 Arrangements should ensure that incompatible chemicals are separated during storage, handling and disposal. At one platemaker's a violent explosion occurred when spent developer containing organic solvents was inadvertently mixed with concentrated nitric acid, a strong oxidising agent. A drum in which the spent developer was being bulked for disposal had not been washed out and still contained some acid.

Bronze and aluminium powders

105 The use and storage of metal powders and pastes containing bronze (copper-zinc) alloys or aluminium can involve a risk of fire due to their finely divided state. These materials may be used in bronzing machines (in powder form) or as a constituent of metallic inks and varnishes. They can react with oxidising agents and chlorinated solvents; in the case of aluminium there are the additional dangers of its reactivity with water, certain other chemicals, and the possibility of creating a dust cloud capable of exploding violently.

106 Where these materials are used, the following precautions are recommended:

(a) the use of powders should be avoided where other suitable alternatives such as pastes or inks are available;

(b) the formation of dust clouds when handling and mixing powders, or by disturbing dried-out paste should be avoided. Suitable extraction may be needed where dust clouds cannot be prevented;

(c) aluminium powder or paste should be kept dry as contact with water can liberate hydrogen and heat;

(d) contact between the powders or pastes and oxidising agents or chlorinated solvents such as 1,1,1-trichloroethane should be avoided;

(e) contact between aluminium powder or paste and alkalis should be avoided;

(f) water or gaseous extinguishers should not be used on a fire involving metal (see Appendix 2, paragraph 7).

Further guidance on aluminum powder and paste is available from the Aluminium Federation (Ref 59).

Printing - litho and letterpress

107 The main problem areas are the use of flammable solvents for cleaning, the use of dryers for drying sheets or webs and the storage of reels at web-fed plant (see paragraphs 74-79).

108 Where possible blanket washes, ink strippers, etc containing solvents with a high flash point should be used; for example, it is normally possible to use a higher flash point solvent instead of methyl ethyl ketone (MEK) which has a flash point of only -5°C.

109 Quantities of flammable liquids in the workroom should be kept to a minimum. They should be kept in a fire-resisting store or cabinet (Figure 9), removed only when required and then returned. The Highly Flammable Liquids and Liquefied Petroleum Gases Regulations 1972 set a maximum of 50 litres when stored in a bin or cupboard in the workroom.

110 Rags impregnated with solvents should be kept in a metal bin with a well-fitting lid (Figure 6). The rags should be removed from the bin to a safe place at least once per day to reduce the risk of spontaneous combustion.

111 Flammable solvents and alcohol for damping solutions should be stored as described in paragraphs 89-92. The small quantities used at machines during a shift should be kept in specifically designed non-spill containers (Figure 5). Open cans and buckets should not be used.

112 Hot air dryers at web-fed presses present a significant hazard of fire and explosion; several explosions have occurred when the web has transferred blanket wash solvent into the dryer and the concentration of solvent within the dryer has exceeded the lower flammable limit. For guidance on minimising the explosion risk and on mitigating the consequences of an explosion see References 48 and 49 and Appendix 6.

113 Where infra-red dryers and ultraviolet curing units are used on machines, automatic means should be provided to shut down the heat/light source if paper remains under the dryer/curing unit for more than a predetermined period.

114 Where paper dust collection systems are used, eg at slitters on large web-fed presses, appropriate explosion relief should be provided (Ref 60).

115 Products should not be stored so as to obstruct means of escape, fire alarm call points, fire-fighting equipment, fire doors or fire shutters.

Printing - flexo and gravure

116 These processes almost always use highly flammable inks, coatings and solvents and they therefore pose serious fire/explosion risks; a fire and explosion involving toluene at a gravure press is described in Appendix 1. The strict requirements of the Highly Flammable Liquids and Liquefied Petroleum Gases Regulations 1972 will apply. Sources of ignition should be avoided and smoking strictly prohibited (see paragraphs 36 and 88 for likely sources of ignition). Wherever possible substitution with less flammable materials should be considered, taking into account any health hazards (Refs 1-3, 24 and 25). Water-based inks should be used where possible.

117 Flexo and gravure areas should be separated as necessary from storage areas and other parts of the building by partitioning having a fire resistance of not less than half an hour.

118 Where mixing has to be done in-house, this should normally be carried out in a special purpose fire-resisting room (separate from the storage area) provided with mechanical ventilation.

119 Where vapour may enter the atmosphere, local exhaust ventilation should be provided to remove the vapour from as close to the source as possible, in order to reduce the risk of fire and explosion and to reduce employee exposure to solvent vapours (see References 1-3, 24 and 25). Examples on a gravure press include the drying hoods and printing stations. The workroom itself should have a good standard of general ventilation.

120 In order to reduce the amount of solvent vapour entering the room and minimise possible spillage, ink ducts, supply containers, etc should be kept covered. It is preferable to pump ink to presses and to return any excess ink/solvent, for example that from ink tanks, to storage via an enclosed system. Where it is necessary to break and remake a connection frequently, for example when making a colour change, a sealed end

coupling is preferred. Where possible pipe runs should be in the open air and suitable shut-off valves should be provided. A manual emergency shut-off valve is recommended where a pipe enters a process building. Pipework and fittings should be to a suitable standard, for example ANSI B 31.3 (Ref 61) and arrangements should be made for routine preventive maintenance including leak testing of pipework, fittings, storage tanks, etc.

121 Highly flammable liquids should not be kept or moved in open-topped vessels. Only properly constructed vessels should be used for mixing and dispensing. Empty or partly used drums of ink or solvent are still hazardous and should not be allowed to accumulate around the press. Highly flammable liquids such as toluene should not be used for floor cleaning.

122 The amount of vapour produced by inks and coatings increases rapidly with temperature so they should be used at as low a temperature as possible. The normal working temperature for an ink or coating should not be exceeded.

123 Any electrical equipment on the press or in the surrounding area should be constructed to a suitable explosion protection standard so as not to present a risk of ignition (see paragraph 88(a)). Powered lift trucks should not be used close to gravure presses or flexo presses unless suitably protected (Ref 54). Where a local exhaust ventilation system is provided for the removal of solvent vapour, the fan motor should not be located in the path of the vapour and the ductwork should be fire-resisting.

124 The generation of static electricity is a problem especially at those presses/laminators/ varnishers which handle plastic or other insulating materials, including many papers. In particular charging of unearthed metal parts as the web passes over cylinders and rollers etc can lead to spark discharges liable to ignite volatile solvents, and people may become charged by induction and by transfer of charge from the web. They may also become charged by walking on an insulating floor or by removing an outer garment. Operations such as pouring, mixing and pumping organic solvents can also generate static electricity. The following precautions should be taken (see also BS 5958: Part 2 (Ref 62) and Refs 63 and 64):

(a) the generation of static electricity from some solvents and other low conductivity liquids should be minimised by avoiding free fall of liquids and restricting pumping speed. For liquids with conductivity up to and including 50 picosiemens per metre (pS/m) the flow velocity in a pipe should not exceed 1 metre per second if a second phase, commonly water, is present. Water may be present even if not deliberately introduced (eg condensate) and so a flow velocity above 1 metre per second should only be considered if there is no possibility of this; only in these circumstances should velocities up to 7 metres per second be considered. Consideration should also be given to the use of anti-static additives to increase conductivity; these reduce the likelihood that a solvent will accumulate a static charge but they will not control static electricity from other sources such as those mentioned above;

(b) vessels, containers, pipework, hoses and plant etc which may become electro-statically charged, either directly or by induction, should be conductive and bonded together and/or earthed. On fixed plant the resistance to earth of all metal or conducting parts should be checked at the commissioning stage and regularly thereafter;

(c) all personnel who may come into contact with a potentially flammable atmosphere should wear anti-static footwear (Refs 65-67); the resistance of footwear while being worn may be measured by means of a personnel resistance monitor. Preferably only outer clothing made from natural fibres should be worn as synthetic fibres can generate static. Although there is no evidence that wearing synthetic underwear can cause a static problem, natural fibres are recommended because injury in the event of fire or explosion is likely to be less severe. Outer clothing such as pullovers and overalls should not be removed in areas where flammable vapours may be present. It should be remembered that even at a level where electrostatic charges cannot be felt, they are capable of igniting some solvent vapours;

(d) floors in hazardous areas (Ref 50) should not be highly insulating; for example concrete would be suitable (see also Reference 68). They should be kept free of insulating deposits;

(e) electrostatic eliminators, of a design incapable of producing incendive sparks should be used on any insulating web-fed material; passive, high voltage and radioactive types are available. This equipment should, where relevant, be constructed to a suitable explosion protection standard so as not to present a risk of ignition (see paragraph 88(a)), and it should be kept clean and properly maintained;

(f) devices for assisting print quality by applying a high electrostatic charge are sometimes used, normally in conjunction with a static eliminator to neutralise the charge before the web moves forward. These devices should be incapable of producing incendive sparks, be constructed to a suitable explosion protection standard (see paragraph 88(a)) and be kept clean and properly maintained;

(g) in order to avoid the possibility of incendive sparks the use of highly insulating plastic materials should be avoided in hazardous areas. In particular powders should not be discharged from plastic bags or liners in the vicinity of flammable atmospheres;

(h) the manual addition of powders or low conductivity liquids to vessels containing a potentially flammable atmosphere should be avoided.

125 Fixed fire detection and extinguishing systems such as carbon dioxide systems designed for both manual and automatic operation are recommended (see paragraphs 57 and 58). When such equipment operates, further flammable solvents should be prevented from entering dryers/extract ducting and, normally, dampers should be fully shut to prevent the extinguishing medium being removed.

126 Particular care should be taken to ensure that dryers incorporate all necessary safeguards to minimise the risk of solvent/air or gas/air explosions, and to mitigate the consequences of an explosion should one occur. For example:

(a) solvent concentrations in dryers and associated ductwork should not exceed 25% of the LFL (see paragraph 84) under all operating conditions (but see sub-paragraph 126(e)). This can normally be achieved by an air flow of 60 m^3 at 16°C for every litre of solvent evaporated. The concentration should be checked by calculation and measurement during commissioning or if the operating conditions are altered. Also, instruments can be used to monitor continuously the solvent concentration in the dryer. The exhaust or inlet ventilation rate, and where appropriate recirculation rate, should be monitored by a differential pressure device or airflow switch. Detection of inadequate air flow should automatically stop the printing process and safely shut down the means of heating. In addition a clear, audible warning should sound automatically. Visible warning may also be provided;

(b) operation of an emergency stop button should safely shut down the means of heating but the exhaust ventilation should continue to operate;

(c) the movement of the printed web (not a dry web) should be possible only if adequate exhaust ventilation has been proved;

(d) adequate explosion relief should be provided on dryers and on associated large-scale ductwork (Refs 48 and 49);

(e) exceptionally some dryers have been designed to operate above 25% of the LFL where there is continuous monitoring. Operation above 25% of the LFL should not be attempted unless the nature of the process ensures that vapour concentrations within the dryer change slowly relative to the effective detection and activation times of the continuous monitor and safety shut-down interlock systems. Under no circumstances should a dryer be operated above a solvent concentration of 50% of the LFL. Such fixed vapour detection equipment should:

(i) be suitable for the solvents to be measured;

(ii) be calibrated for the solvents concerned, regularly tested and recalibrated, and well maintained;

(iii) normally sample at points within the dryer and/or ductwork where the vapour concentration is likely to be highest;

(iv) operate an audible and visible alarm if the solvent vapour concentration at separate sampling points exceeds the normal operating limit;

(v) have two independent reliable instruments measuring the solvent concentration at separate sampling points, each of which can safely shut down the printing process and the dryer's means of heating before the solvent vapour concentration rises above a predetermined maximum (which should not be above 50% of the LFL);

(vi) be arranged to open fully any modulated dampers in the event of a malfunction in the continuous monitoring system;

(vii) automatically and continuously monitor and record the progression of solvent vapour concentration with time.

Information on the use and limitations of flammable vapour detectors can be found in Reference 69.

(f) where there is a solvent recovery unit, a damper should be provided in the dryer extract duct to isolate the unit from any fire on the press or in the ducting. Appendix 5 describes the precautions required at solvent recovery units operating by condensation.

127 Vapour detection equipment is sometimes provided near printing units in order to detect escapes of solvent into the general atmosphere of the workroom. Where such equipment is not provided, regular routine measurements should be considered to ensure that the atmosphere is maintained below 25% of the LFL. One method of doing this would be to use a portable flammable gas detector to check for escapes of vapour at flanges, valves, pump seals and other potential sources of leaks.

128 Where appropriate, automatic viscosity measuring equipment should be provided on the machine or it should be possible to take samples from a position at the side of the machine rather than from between the units.

129 Where highly flammable liquids are used to clean rollers, cylinders and ancillary equipment, the operation should preferably be done in a proprietary solvent-cleaning machine fitted with exhaust ventilation. In other cases cleaning should take place in a purpose-designed booth; the enclosure and extract ducting should be fire-resisting and the electric motor driving the extraction fan should not be in the path of the vapour. Where rollers/cylinders are cleaned by hand on the press, only small volumes should be applied in well-ventilated conditions. The solvents should be kept in non-spill containers.

Printing - screen

130 In general inks, thinners, retarders, etc used in screen printing are flammable and therefore need to be controlled carefully as described in paragraphs 83-94. Water-based inks should be used where possible.

131 A good standard of housekeeping should be maintained. Workrooms should be kept clean and tidy and waste bins emptied regularly; used rags or wipers should be stored in a metal bin with a well-fitting lid (Figure 6).

132 Solvent concentrations in hot air dryers and their associated ductwork should be kept below 25% of the LFL and be discharged to a safe place in the atmosphere. Where the ductwork is long, drain points should be provided for removal of condensates.

133 Free standing multi-rack dryers should be located well away from sources of ignition in an area with a high standard of ventilation.

Bindery and finishing

134 This includes cutting, folding, gathering, trimming, stitching, sewing, case making, gold blocking, laminating, waxing or other coating, slitting, packing and despatch.

135 The main problem in these areas is the amount of loose paper produced, especially at guillotines and trimmers. Suitable receptacles, preferably non-combustible, should be provided

near to the machine. Waste should be removed from the workroom as soon as possible and then baled or otherwise disposed of. Where appropriate, waste should be removed automatically either by conveyor or pneumatic systems.

136 Products should not be stored so as to obstruct means of escape, fire alarm call points, fire detection equipment, fire-fighting equipment, fire doors or fire shutters.

137 Floors and machines should be cleaned of waste at least once a shift with particular attention paid to the waste which may fall within or behind machines.

138 Heat sources within bindery and finishing machines should be designed and constructed to prevent ignition of materials. Temperature controls on gas and electrical heating equipment should have back-up protection as a precaution against the failure of the primary control, eg a thermostat or thermal cut-out operating at a temperature a few degrees higher than the normal maximum operating temperature. Appropriate interlocking arrangements should be provided, for example heating should be switched off when a machine stops to prevent the process material becoming overheated.

139 Piped supplies of flammable materials, eg some adhesives and varnishes, should preferably be through rigid metal pipes, or other material designed to a suitable standard. Surfaces in the vicinity of a coating process should be preferably non-absorbent and any deposits cleaned regularly. Spillages should be cleaned as soon as they occur.

140 Heating elements or flames should be so positioned or guarded that ignition of the product or material being processed is prevented if the paper buckles or its pages fall from a conveyor. Where highly flammable adhesives are used, adequate mechanical exhaust ventilation should be provided.

141 A dust extraction and collection system should be provided for slitters, paper roughing devices, book saws, etc with appropriate explosion relief where necessary (Ref 60).

142 Glue pots heated by gas or electricity should have non-combustible, thermally insulated stands. Electric pots should be fitted with 'mains on' warning indicator lamps and should be provided with thermostatic controls including an upper limit safety cut-out with manual reset.

SUMMARY

143 People have been killed or injured by fires and explosions in the printing industry. Many of the materials used are combustible and every year there are several hundred fires requiring the attendance of the fire service and causing considerable financial loss.

144 To help reduce this toll it is vital that senior managers are committed to preventing fires and explosions and ensuring that workers are not put at risk should either occur. Fire and explosion risks require evaluation backed up by effective control. Means of escape need to be identified and kept free of obstruction; exits need to be readily openable from the inside.

145 People need to be adequately trained so that they are able to play their part in minimising fire risks and know what to do if fire breaks out or other emergency occurs. Key staff should be fully involved and safety representatives consulted.

146 The aim is to stop a fire starting, but if one does occur, to enable people to escape safely and to prevent it spreading. Day-to-day control is essential to keep the quantity of flammables to a minimum, to ensure safe handling of flammable liquids, to exclude sources of ignition and to keep up standards of housekeeping.

147 Everyone needs to be vigilant in their own interest and that of other people.

APPENDIX 1: EXAMPLES OF FIRES IN THE PRINTING INDUSTRY

1 Many of the following incidents in the printing industry resulted in death and injury. Financial loss in some cases amounted to several million pounds.

Fire in reel store

2 Three employees died in a fire in the reel store of a printing factory. The fire started in some reels of paper at one end of the reel store in this single-storey factory. There was no fire separation between the reel storage area and the production area. Shortly after the fire was first discovered there was a sudden rapid spread of fire (flashover) across the top of reels of paper and the whole building rapidly filled with smoke. The roof collapsed within minutes.

3 Of 42 people in the building at the time of the fire, 40 managed to escape when the alarm was sounded. Two people who were working in a small row of rooms lost their lives, together with another employee who had returned to the building to try to help them escape. Their bodies were found beneath collapsed paper and debris not far from an exit.

4 The reels in the store were stacked on end, commonly to a height of six reels. The reels where the fire started consisted of partially used reels covered in plastic film and damaged reels which were easily ignitable. The reels in the main part of the store were wrapped in plastic film or tar impregnated paper. The sideways spread of fire and heat was fanned by high winds through an open door and this would have been aided by the relatively small gap between the tops of the reels and the underside of the roof.

5 The row of small internal rooms was at one side of the reel store. A corridor was formed by the rooms and the wall of stacked paper reels with its potential for rapid surface flame spread.

6 There were six well-distributed exits from the factory and a fire alarm and fire-fighting equipment were provided. There was apparently no smoke venting in the roof, no sprinkler system and the smoke detector beams were turned off at the time of the fire.

7 Forensic investigators were unable to determine the cause of the fire.

Fire and explosion at gravure printing press

8 One employee was burnt, another was affected by smoke inhalation, and they and a number of other employees suffered from shock as a result of a fire and explosion at a gravure printing press involving toluene. A neighbouring gravure press linked by ducting associated with a solvent recovery unit caught fire and the force of the explosion pushed out a noise wall, blew off roof panels and badly damaged a blockwork wall. Parts of the solvent recovery unit were also ruptured and distorted. The seat of the explosion was thought to be within the extract ducting.

9 The drying system comprised a steam-heated hood above each print unit through which the web passed after emerging from the printing cylinders. A riser duct containing a damper and sampling point connected each hood to a main horizontal duct carrying vapours out of the press hall to the solvent recovery unit. Gas analysers sampled the air in the riser ducts for toluene vapour. The analysers fed a control system which varied a damper in each riser duct so as to maintain a preset toluene vapour concentration slightly above 25% of the lower flammable limit.

10 The press was running on slow when there was a whoosh of flame followed seconds later by a violent explosion. A definite cause could not be ascertained but an increase in vapour concentration may have coincided with a fire caused by static electricity.

11 Among the changes made following the incident were the installation of a modified vapour detection system (including sampling points between press units and above each press), better control of static electricity and improvements to the handling and use of toluene. The extract ducting was separated to create individual extraction units for each press and thereby avoid transfer of any ignition from one system to the other.

Faulty electrical appliance

12 The premises of a printing and book binding firm were severely damaged by a fire resulting from a fault developing on an adhesive heating

appliance. Overheating resulted in the production of flammable vapours from resin-based adhesive which were in turn ignited by hot surfaces.

Accumulation of paper in lift truck engine compartment

13 The roof and walls of a paper store were totally destroyed by a fire which also consumed some 1500 tonnes of paper reels, a large quantity of magazines and timber pallets. A lift truck, a tractor unit and a trailer were also severely damaged. Only the action of the fire brigade saved the adjacent print room.

14 A lift truck driver had earlier been transporting reels of paper from the paper store to the print room. He parked the truck in the centre of the store near reels of paper stacked on end, switched off the engine and walked into the print room. On returning to the store approximately four minutes later, he found the lift truck engulfed in flames. The fire had already spread to the reels of paper and reached roof level.

15 The fire brigade was called. Employees attempted to tackle the fire using portable extinguishers and a jet from a fixed internal hose reel, but by then the fire was reaching severe proportions and spreading rapidly at roof level; they were forced to withdraw and the factory was evacuated.

16 Investigation showed that the hot exhaust manifold of the lift truck had ignited paper which had accumulated in the engine compartment.

Defective dryer

17 A printing room suffered considerable damage when a fire occurred in an infra-red dryer attached to a lithographic printing press. When the transportation mechanism failed, paper remaining in the dryer overheated, ignited and involved further quantities of paper.

18 Smoke damaged some of the printing machinery and contents of the printing room. There was also smoke damage in other parts of the building. The drying machine itself was severely damaged as were several metres of extraction ducting.

Arson

19 On arriving at a warehouse just after midnight the fire brigade found a timber tea chest or pallet on fire outside double timber doors in an alleyway adjacent to the seven-storey building. The doors were burning and fire was spreading under the doors into a stairwell. The first and second floors were also alight. Fire spread was rapid due to the type of storage and unprotected stairways in the warehouse areas. The entire seven-storey warehouse was severely damaged.

Ignition of toluene vapour

20 A printer was off work for six weeks after receiving severe burns to his right hand when a static discharge occurred as he was using a toluene soaked rag to clean the doctor blade of a gravure printing press. It is thought that an earlier modification to a cover at the doctor blade had reduced the extraction system's ability to remove toluene vapours given off by the printing ink. Anti-static footwear was not in use.

Smoking near flammable solvent

21 An employee sustained burns to his face and hair as he cleaned a roller washing machine with flammable solvent while smoking. After the incident all staff were instructed not to smoke in the room where the machine was located and 'no smoking' signs were posted up.

Horseplay with flammable thinners

22 A 17-year-old apprentice died of extensive burns after an initiation ceremony in which 2 litres of flammable thinners were poured on his clothing. The victim was changing his soaked clothing in a toilet cubicle when some matches were lit. The apprentice was engulfed in flames and died three weeks later of burns which covered 80% of his body.

Ignition of paper dust accumulations

23 A serious fire occurred at the folder of a web offset printing press causing damage to electrical cables, lubricant feed pipes etc. A suspected static discharge produced a flash which initially ignited paper dust inside the folder. The fire spread rapidly through dust accumulations to the paper web. The

blaze was controlled by employees using fire extinguishers. The company decided to install a CO_2 gas flooding system and static eliminators and to improve cleaning routines.

Toluene leak

24 Some 1200 litres of toluene leaked into a basement electrical control room from a domestic water-type compression joint in a copper pipe supplying a gravure printing press. Two days earlier, the nine-year-old system had been converted from gravity feed to pressure feed at 2 bar. Fortunately there was no ignition and no injuries. The fire brigade was called and dealt with the leak using sand and a high velocity fan. No design specification had been provided for the pipework and there had been no routine maintenance. The pipework was subsequently replaced in steel with threaded joints.

APPENDIX 2: SELECTING THE APPROPRIATE EXTINGUISHER

1 There are four main classes of fire - A, B, C and D.

Class A fires

2 **Fires involving solid materials, usually of an organic nature, eg wood or paper in which combustion normally takes place with the formation of glowing embers.** Class A fires occur frequently and it will be appropriate to provide fire-fighting equipment suitable for this class of fire. Water, foam (other than protein foam) and multi-purpose powder are the most effective media for extinguishing these fires. Water and foam are usually considered the most suitable media, and the appropriate equipment would therefore be hose reels, water-type extinguishers or extinguishers containing fluoroprotein foam (FP), aqueous film forming foam (AFFF) or film forming fluoroprotein foam (FFFP).

3 If hose reels are installed, they should be located where they are conspicuous and always accessible. Their distribution should be such that, with not more than 45 metres of hose, no part of the area to be protected is more than 6 metres from the nozzle of the reel when the hose is fully run out. Hose reel installations should conform with the appropriate recommendations in British Standards 5306: Part 1 (Ref 31) and 5274 (Ref 70).

4 If portable fire extinguishers are installed, they should be provided and allocated to accord with the recommendations contained in British Standard 5306: Part 3: clause 5.2 which deals with extinguishers suitable for Class A fires (Ref 31).

Class B fires

5 **Fires involving flammable liquids or liquefiable solids (greases and fats).** In buildings or parts of buildings where there is a risk of fire involving flammable liquid it will usually be appropriate to provide portable fire extinguishers of foam, carbon dioxide (CO_2) or powder types. Care should be taken when using gaseous extinguishers as the fumes and products of combustion may be hazardous in confined spaces. For environmental reasons it is recommended that the provision of halon extinguishers should be avoided where other suitable extinguishing media are available. Table 1 of clause 5.3 of British Standard 5306: Part 3 (Ref 31) gives guidance on the minimum scale of provision of various extinguishing media for dealing with a fire involving exposed surfaces of contained liquid.

Class C fires

6 **Fires involving gases.** No special extinguishers are made for dealing with fires involving gases. While dry powder extinguishers are capable of putting out small gas fires, normally the only effective action against such fires is to stop the flow of gas. Indeed there would be a risk of an explosion if a fire involving escaping gas were to be extinguished before the supply could be cut off.

Class D fires

7 **Fires involving metals.** None of the extinguishing agents referred to in the preceding paragraphs will effectively deal with a fire involving finely divided metals. These fires should be dealt with only by trained personnel using special extinguishers.

Fires involving live electrical equipment

8 Extinguishers provided specifically for the protection of electrical risks should be of the dry powder or CO_2 type. While some extinguishers containing aqueous solutions such as AFFF may meet the requirements of the electrical conductivity test of BS 5423 (Ref 32), they may not sufficiently reduce the danger of conductivity along wetted surfaces such as the floor. Consequently such extinguishers should not be provided specifically for the protection of electrical risks and where such extinguishers are provided in close proximity to electrical equipment, precautions should be taken to switch off the electrical supply before application.

General

9 Fire extinguishers should conform to British Standard 5423 (Ref 32). Those which do are approved by the British Standards Institution and certified under the British Approvals for Fire Equipment (BAFE) scheme. Conforming equipment bears the LPC (Loss Prevention

Council) or BAFE marks of approval. All extinguishers should be installed and maintained as outlined in BS 5306: Part 3 (Ref 31). It is recommended that the contract maintenance of portable fire extinguishers is entrusted to a competent person, such as a BAFE registered firm.

10 In some circumstances it may be more appropriate to use a fire blanket rather than an extinguisher, eg to deal with a glue pot fire in the early stages, or to smother a fire involving a person's clothing. Such blankets should be to BS 6575 (Ref 71) and be asbestos free.

Fire extinguishers

WATER

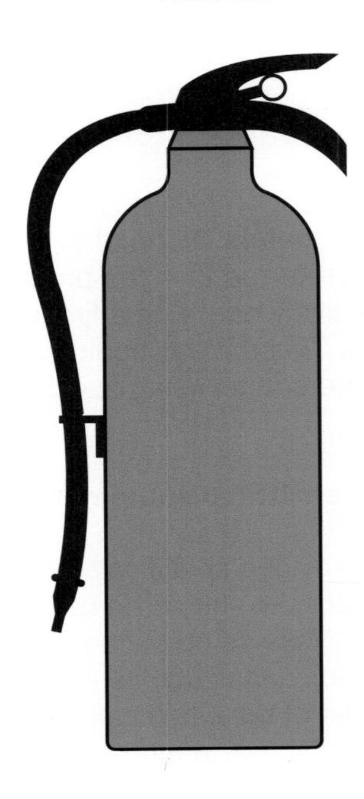

RED

EXTINGUISHING ACTION

Mainly by cooling the burning material.

CLASS OF FIRE

Class A

DANGER Do not use on live electrical equipment, burning fats or oils.

METHOD OF USE

The jet should be directed at the base of the flames and kept moving across the area of the fire. Any hot spots should be sought out after the main fire is out.

FOAM (Protein P) Type

CREAM

EXTINGUISHING ACTION

Forms a blanket of foam over the surface of the burning liquid and smothers the fire.

CLASS OF FIRE

Class B

DANGER Do not use on live electrical equipment.

METHOD OF USE

The jet should not be aimed directly onto the liquid. Where the liquid on fire is in a container the jet should be directed at the edge of the container or on a nearby surface above the burning liquid. The foam should be allowed to build up so that it flows across the liquid.

AQUEOUS FILM FORMING FOAM (AFFF)

Film-forming Fluoroprotein foam (FFFP)
Fluoroprotein foam (FP)

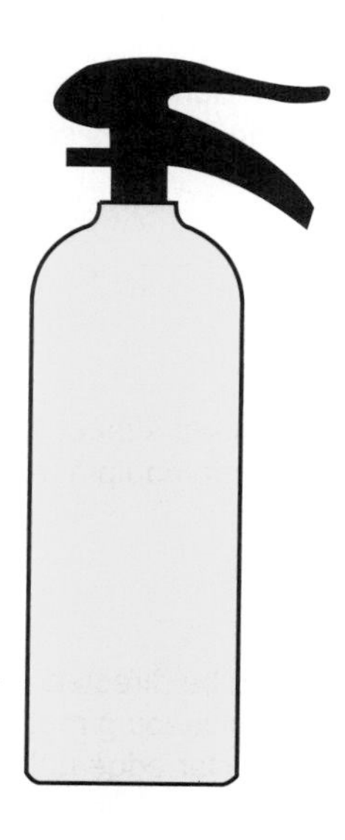

CREAM

EXTINGUISHING ACTION

Forms a fire extinguishing water film on the surface of the burning liquid. Has a cooling action with a wider extinguishing application than water on solid combustible materials.

CLASS OF FIRE

Classes A and B

DANGER Some extinguishers of this type are not suitable for use on live electrical equipment.

METHOD OF USE

For Class A fires the directions for water extinguishers should be followed.

For Class B fires the directions for foam extinguishers should be followed.

POWDER

BLUE

EXTINGUISHING ACTION

Knocks down flames.

CLASS OF FIRE

Class B

Safe on live electrical equipment although does not readily penetrate spaces inside equipment. A fire may re-ignite.

METHOD OF USE

The discharge nozzle should be directed at the base of the flames and with a rapid sweeping motion the flame should be driven towards the far edge until the flames are out. If the extinguisher has a shut-off control the air should then be allowed to clear; if the flames reappear the procedure should be repeated.

WARNING Powder has a limited cooling effect and care should be taken to ensure the fire does not re-ignite.

POWDER (Multi-purpose)

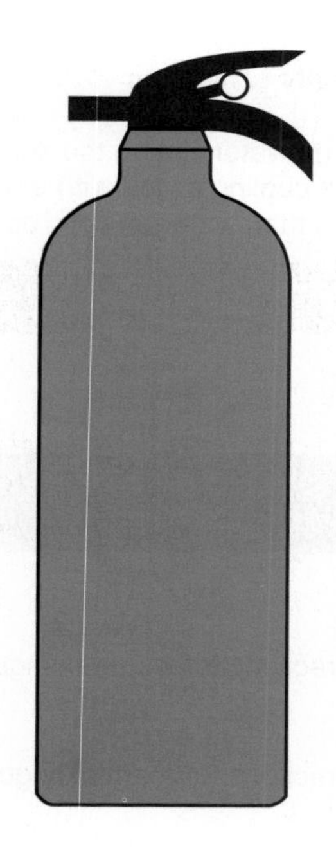

BLUE

EXTINGUISHING ACTION

Knocks down flames and on burning solids melts down to form a skin smothering the fire. Has some cooling effect.

CLASS OF FIRE

Classes A and B

Safe on live electrical equipment although does not readily penetrate spaces inside equipment. A fire may re-ignite.

METHOD OF USE

The discharge nozzle should be directed at the base of the flames and with a rapid sweeping motion the flame should be driven towards the far edge until the flames are out. If the extinguisher has a shut-off control the air should then be allowed to clear; if the flames reappear the procedure should be repeated.

WARNING Powder has a limited cooling effect and care should be taken to ensure the fire does not re-ignite.

CARBON DIOXIDE (CO_2)

BLACK

EXTINGUISHING ACTION

Vaporising liquid gas which smothers flames by displacement of oxygen in the air.

CLASS OF FIRE

Class B

Safe and clean to use on live electrical equipment.

METHOD OF USE

The discharge horn should be directed at the base of the flames and the jet kept moving across the area of the fire.

WARNING CO_2 has a limited cooling effect and care should be taken to ensure that the fire does not re-ignite.

DANGER Fumes from CO_2 extinguishers can be harmful to users in confined spaces. The area should therefore be ventilated as soon as the fire has been extinguished.

HOSE REEL

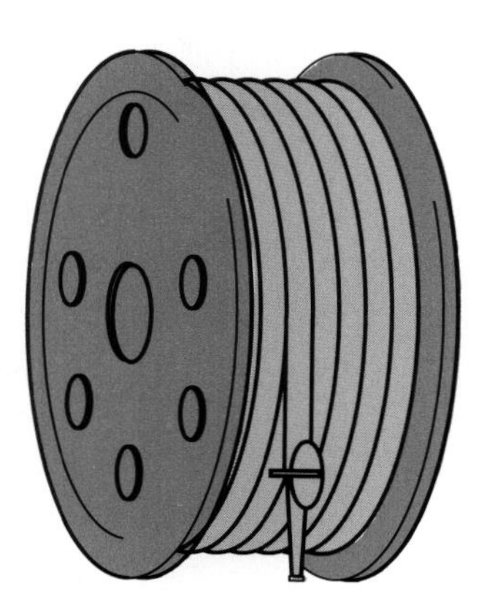

RED

EXTINGUISHING ACTION

Mainly by cooling the burning material.

CLASS OF FIRE

Class A

DANGER Do not use on live electrical equipment.

METHOD OF USE

The jet should be aimed at the base of the flames and kept moving across the area of the fire. If an isolating valve is fitted it should be opened before the hose is unreeled.

FIRE BLANKET

RED

EXTINGUISHING ACTION

Smothering

CLASS OF FIRE

Classes A and B

Light duty - Suitable for burning clothing and small fires involving burning liquids.

Heavy duty - Suitable for industrial use. Resistant to penetration by molten materials.

METHOD OF USE

The blanket should be placed carefully over the fire and the hands shielded from the fire. Care should be taken that the flames are not wafted towards the user or bystanders.

British Standard 5423 recommends that extinguishers should be (a) predominantly red with a colour coded area; (b) predominantly coloured coded; or (c) of self-coloured metal with a colour coded area.

APPENDIX 3: PERMIT-TO-WORK

1 Certain activities, eg the use of welding, flame cutting or portable grinding equipment (particularly in areas where paper or flammable liquids are stored or used), need to be strictly controlled and this can best be done through a written permit-to-work for the people involved. This should apply both to in-house and contractors' staff.

2 Any employees and outside contractors who are to take part in such activities should be left in no doubt that work may not begin until the person who is to issue the permit has explained fully the safety precautions that must be observed. It is imperative that a written handover procedure is adopted.

3 In operating a permit-to-work system the following principles should be observed:

(a) the information given in the permit should be clear and unambiguous;

(b) it should specify precisely and in detail the item of plant on which work is to be carried out, the nature of the operations, the point at which welding or hot work is to take place and the precautions which should be taken to ensure safety of personnel;

(c) the permit should specify the time at which it comes into operation, the time by which it expires and any particular conditions under which all work should cease even though the time limit for the certificate has not expired;

(d) the person issuing the permit should be satisfied by personal inspection that all the action specified as necessary has in fact been taken;

(e) the person issuing the certificate should have the technical knowledge not only to appreciate the existence of hazards and the precautions to be taken, but also the authority to require responsible people to make safety recommendations on matters of which they have special knowledge and to co-ordinate the duties of all concerned.

4 The permit-to-work procedure outlined in paragraphs 1-3 is for general guidance and should be adapted to suit particular needs. An example is given overleaf.

HOT WORK PERMIT

Applicable to all operations involving flame, hot air or arc welding and cutting equipment, brazing and soldering equipment, blowlamps, bitumen boilers and other equipment producing heat or having naked flames.

PART 1

Date: ....

Permission is granted to: ....

to use .... in the .... (exact location)

between .... am and .... pm

.... pm and .... pm

I have examined the above location.

There are no combustible liquids, vapours, gases or dusts.

All combustible material has either been removed or suitably protected against heat and sparks.

A person is to stand by with an extinguisher/hose reel while the operation is in progress.

Those concerned have had the nearest fire alarm/telephone pointed out to them and have been told what to do in the event of a fire.

Signature of person issuing permit and position held ....

PART 2

Work has been completed and all sources of ignition removed. The work area and all adjacent areas to which sparks and heat might have spread were thoroughly inspected on completion of the operation and at least one hour later in order to ascertain that no smouldering fires were discovered.

Signature of person responsible for the work ....

PART 3

The location is safe and flammable processes/storage may be resumed. This hot work permit is now cancelled.

Signature of person issuing permit ....

APPENDIX 4: FIRE-RESISTING STRUCTURES

1 For storerooms, workrooms, cupboards, bins, process enclosures, ducts, etc which are required to be fire-resisting under the Highly Flammable Liquids and Liquefied Petroleum Gases Regulations 1972, Certificate of Approval No 1 has been issued by HM Chief Inspector of Factories. This should form the basis of construction of fire-resisting storerooms etc whether or not the specific Regulations apply. The certificate itself, form F2434, is currently out of print and under revision. Until it is reissued the following criteria for fire resistance should be used.

External storage buildings and internal storerooms

2(a) Each enclosing element, eg walls, doors, windows, floors and ceilings, should be at least half-hour fire resistant to BS 476 Part 8 (now replaced by Parts 20-23) - see Reference 72. Exceptions are floors immediately above the ground, and tops or ceilings of single-storey buildings and of top-floor rooms. For doors the insulation requirement is waived. No glazing is allowed in internal walls forming part of the storeroom enclosure.

(b) All internal surfaces should be at least Class 1 if tested to BS 476 Part 7 (surface spread of flame). Exceptions are floors, and doors, windows and their frames.

(c) Doors should be self-closing from any position except doors which lead directly out of the building.

(d) Junctions between elements of construction should be bonded or fire-stopped to prevent or retard the passage of flame and hot gases.

(e) The structure should be robust enough to withstand foreseeable accidental damage.

(f) If surfaces of the structure are liable to be coated with residues, the structure should allow removal of the residues without affecting its fire resistance or its ability to resist flame spread.

Workrooms

Note: The following requirements do not apply to external doors, windows and walls, ventilation openings or to tops or ceilings of single-storey buildings and top-floor rooms.

3(a) Each enclosing element, eg walls, doors, windows, floors and ceilings, should be at least half-hour fire resistant to BS 476 Part 8 (now replaced by Parts 20-23) - see Reference 72. An exception is floors immediately above the ground. For floors, ceilings or doors the insulation requirement is waived.

(b) All internal surfaces should be at least Class 1 if tested to BS 476 Part 7 (surface spread of flame). Exceptions are floors, and doors, windows and their frames.

(c) Doors should be self-closing from any position.

(d) Junctions between elements of construction should be bonded or fire-stopped to prevent or retard the passage of flame and hot gases.

(e) The structure should be robust enough to withstand foreseeable accidental damage.

(f) If surfaces of the structure are liable to be coated with residues, the structure should allow removal of the residues without affecting its fire resistance or its ability to resist flame spread.

Cabinets and enclosures

4(a) In the case of a cabinet or enclosure (other than an oven used solely for the evaporation of flammable solvents from materials inside the oven or a fume cabinet or glove box):

(i) each side, top, floor and door should be half-hour fire resistant to BS 476 Part 8 (now replaced by Parts 20 and 22);

(ii) the internal surface material (and any substrate to which it is bonded) should, if tested to BS 476 Part 6 (fire propagation test), have an index of performance of not more than 12 and a sub-index of not more than 6.

(b) Each side, top, floor and door of ovens used only for the evaporation of flammable solvents from materials inside the oven should be non-combustible if tested to BS 476 Part 4 or, if tested to BS 476 Part 11, should not flame and should give no rise in temperature on either the centre (specimen) or furnace thermocouples.

(c) Each side, top, floor and door should be supported and fastened to prevent failure of the structure in a fire for at least half an hour. Supports and fastenings should be non-combustible to BS 476 Part 4 or, if tested to BS 476 Part 11, should not flame and should give no rise in temperature on either the centre (specimen) or furnace thermocouples. Exceptions are fume cabinets and glove boxes.

(d) In the case of every cabinet or enclosure:

(i) junctions between sides, tops and floors should be bonded or fire-stopped to prevent or retard the passage of flame and hot gases;

(ii) the structure should be robust enough to withstand foreseeable accidental damage;

(iii) if surfaces of the structure are liable to be coated with residues, the structure should allow removal of the residues without affecting its fire resistance or its ability to resist flame spread.

Cupboards and bins

5(a) Each side, top, floor and lid should be able to satisfy BS 476 Part 8 (now replaced by Parts 20 and 22) with regard to freedom from collapse and resistance to passage of flame for at least half an hour.

(b) The internal surface material (and any substrate to which it is bonded) should, if tested to BS 476 Part 6 (fire propagation test), have an index of performance of not more than 12 and a sub-index of not more than 6.

(c) Junctions between each side, top and floor should be bonded or fire-stopped to prevent or retard the passage of flame and hot gases.

(d) The structure should be robust enough to withstand foreseeable accidental damage.

(e) If surfaces of the structure are liable to be coated with residues, the structure should allow removal of the residues without affecting its fire resistance.

(f) Each side, top and floor should be supported and fastened to prevent failure of the structure in a fire for at least half an hour. Supports and fastenings should be non-combustible if tested to BS 476 Part 4 or, if tested to BS 476 Part 11, should not flame and should give no rise in temperature on either the centre (specimen) or furnace thermocouples.

Ducts, trunks and casings

6(a) Each duct, trunk and casing should be able to satisfy BS 476 Part 8 (now replaced by Parts 20 and 22) with regard to freedom from collapse and resistance to passage of flame for at least half an hour.

(b) The internal surface material (and any substrate to which it is bonded) should, if tested to BS 476 Part 6 (fire propagation test), have an index of performance of not more than 12 and a sub-index of not more than 6.

(c) The structure should be robust enough to withstand foreseeable accidental damage.

(d) If surfaces of the structure are liable to be coated with residues, the structure should allow removal of the residues without affecting its fire resistance.

(e) Each duct, trunk and casing should be supported and fastened to prevent failure of the structure in case of internal fire for at least half an hour. Supports and fastenings should be non-combustible to BS 476 Part 4 or, if tested to BS 476 Part 11, should not flame and should give no rise in temperature on either the centre (specimen) or furnace thermocouples.

APPENDIX 5: SOLVENT RECOVERY

Where solvent recovery units operating by condensation are used, the following precautions are necessary (there are also hazards associated with the operation of carbon adsorption beds about which specialist advice should be sought, for example from firms having the appropriate expertise):

(a) the plant should be sited with good natural ventilation, preferably outside in the open air, on an impervious non-combustible base with a spillage retention sill;

(b) the use of flexible connections and hoses should be kept to an absolute minimum. All materials used in the construction should be resistant to the solvent and other materials passing through the process;

(c) high level alarms or some other means of high level warning should be installed on the distillation vessel and receiving vessel where any possibility of overfilling might occur, eg pumped systems or where the receiving vessel is or can be smaller than the batch volume. Only in the simplest cases in the open air will dipping be acceptable;

(d) the operating pressure of the system should not exceed its designed working pressure. If there is a likelihood of the condenser becoming blocked, the vessel should be fitted with a pressure relief device which vents to a safe place in the open air;

(e) still temperatures should be substantially below the auto-ignition temperature of any material in the vessel;

(f) several serious incidents have arisen from the failure to condense flammable vapours and the control and monitoring system associated with the condenser should therefore be carefully examined. Positive flow monitoring of the water flow in water-cooled systems is preferred, but monitoring of the exit vapour or liquid temperature is acceptable. A warning alarm should sound, and the plant (except for the condenser cooling supply) should shut down if any serious malfunction occurs, eg excess temperature rise of cooling water or control power supply failure. The condenser should be fitted with a vent at its outlet. The vent should discharge to a safe place in the open air;

(g) no electrical equipment should be in the path of the pressure relief device or near the vent discharge point. Areas within 1.5 metres of the vent pipe or where there is a likelihood of flammable vapours being present during normal operation, such as loading points, should be classified as Zone 1 (Ref 50). All other areas within 2 metres of the plant should be classified as Zone 2;

(h) precautions should be taken to reduce the static ignition hazard. These include the bonding and earthing of all parts of the plant, including ancillary vessels etc, and appropriate operator instruction (see paragraph 124);

(i) distillation residues are flammable and capable of evolving the vapours of all the original mixture components. If they are handled hot and as a liquid, they should be removed via a closed system to a suitable container. If they are to be handled cold and as a solid or semi-solid, the still should be allowed to cool sufficiently for the residues to evolve minimal vapour when the still is opened. The toxic risks to operators in both cases must be considered carefully and appropriate precautions taken (see Reference 73);

(j) all dirty and recovered solvents and flammable liquid residues should be clearly labelled and stored, where appropriate, in accordance with the Classification, Packaging and Labelling of Dangerous Substances Regulations 1984, or the advice given in References 55-57 or by the local petroleum licensing authority.

APPENDIX 6: SAFETY PRECAUTIONS FOR BLANKET WASH SYSTEMS IN HEAT-SET WEB OFFSET PRINTING

The hazards

1 A number of explosions have occurred in dryers on web offset printing lines when the web has transferred blanket wash solvent into the dryer and the concentration of solvent within the dryer has exceeded the lower flammable limit (LFL) - see paragraph 84 on page 15.

2 Such explosions can be very violent and have the potential to cause serious injury, as well as damage to property. In one instance the force of the explosion badly distorted the dryer, displaced a substantial factory wall and broke nearby windows. Printers in the vicinity were fortunate to escape serious injury.

3 The design and operation of heat set dryers are such that ignition of solvent vapour is almost inevitable if a concentration above the LFL is formed within the dryers. Precautionary measures to prevent the formation of flammable concentrations of solvent within dryers and to mitigate the consequences of any explosion are essential and are set out in the following paragraphs. These standards are based on experience to date and have been agreed by the relevant printing employers trade associations, the printing trade unions and by the major suppliers of dryers and automatic blanket wash systems. As most heat set dryers are direct gas fired, gas safety precautions are also included.

Preventive measures

4 Where blanket washing is undertaken *by hand*, the web should be removed or broken before the dryer and should not be used as a wiper. In addition paragraphs 7-12, 14 and 26 of this appendix apply. (The quantity of blanket wash solvent applied should, of course, be kept to the minimum.)

5 All installations with *automatic* blanket wash systems should comply with paragraphs 7 to 26 of this appendix. Users should consult with the suppliers of their plant and equipment and check the safety features incorporated in existing installations.

6 Suppliers of heat-set dryers, automatic blanket wash systems and presses should co-operate together and with users of their equipment to exchange relevant information and so enable their equipment and plant to be designed, installed, commissioned and run safely. Particular care is required in the design and operation of automatic blanket wash systems in which the print units are washed simultaneously rather than one at a time.

Dryers

Gas safety

7 Gas safety features, including safety shut-off valves, ignition systems, flame failure devices and purge times should be in accordance with BS 5885 *Automatic gas burners* Parts 1 and 2 (Ref 74) or with British Gas publication IM18 *Code of practice for the use of gas in low temperature plant* (Ref 75) as appropriate.

Explosion relief

8 Adequate explosion relief doors or panels should be provided. The size of the explosion relief should be estimated according to one of the following techniques:

(a) an area of relief equal to at least half the area of one long side for dryers of square or approximately square cross-section. For other dryers the minimum relief should be taken to be half the area of the top of the dryer or half the area of one long side, whichever is the greater. Where existing dryers do not meet this standard, the area of relief should be increased so as to be as close to it as possible;

(b) an area of relief calculated using the vent ratio method. This requires an area of explosion relief in square metres based on a ratio of 1:4.5 of the dryer volume in cubic metres (1:15 in imperial units). There are certain limitations with this method;

(c) an area of relief calculated using an empirical method when the strength of the dryer and the solvent burning characteristics are known. Such methods detailed in NFPA 68 (Ref 76) and in a review by Marshall (Ref 77) are acceptable.

9 The weight of explosion relief doors or panels should be as low as possible and should be no greater than 25 kg/m^2 (5 lb/ft^2). In addition the opening pressure of the relief should be as low as possible and should not exceed 250 kgf/m^2 (50 lbf/ft^2). The opening pressure should not be capable of unauthorised adjustment.

10 Explosion relief doors or panels should be distributed uniformly along the length of the dryer and should vent to a safe place away from personnel or flammable materials. If, in the event of an explosion, the explosion reliefs could become dangerous missiles they should be chained or otherwise restrained. Venting of explosion reliefs should not be obstructed by pipework or other fixtures. Where explosion reliefs cannot be located on the top of the dryers, barriers or deflector plates may be necessary to protect personnel from the effects of an explosion. Dryer doors not used as explosion relief should be adequately secured.

11 Further guidance on explosion relief is available in HSE booklets HS(G)16 (Ref 48) and HS(G)11 (Ref 49).

Exhaust ventilation

12 Sufficient exhaust ventilation should be maintained from the dryer to ensure that under all operating conditions the concentration of solvent in the dryer does not exceed 25% of the LFL. The exhaust ventilation rate should be monitored by a differential pressure switch or similar device in the exhaust duct or air inlet. Detection of inadequate air flow should result in automatic safe shut-down of the fuel supply to the dryer and of the automatic blanket wash system as appropriate, and should activate a visible/audible alarm. Any damper in the exhaust duct should be cut away so that at least one third of the cross-sectional area of the duct remains open even when the damper is fully closed. Alternatively a stop may be fitted on the damper movement to achieve the same result.

13 Where an automatic blanket wash system is used, the solvent loading in the dryer will increase. Automatic means should be provided to ensure that in all circumstances the overall concentration of solvent in the dryer does not exceed 25% of the LFL* (see also paragraph 18). As this may require an increased rate of exhaust ventilation, automatic means should be provided to open fully the exhaust damper and, if necessary, to increase the speed of the extraction fan before and during the blanket wash cycle. The increased rate of exhaust ventilation should be proved, eg by means of a second differential pressure device or by a position switch on the damper in the exhaust duct.

14 Adequate written information should be provided to users about the purposes for which the dryer has been designed and tested, the conditions necessary for safe use, the safety features incorporated and the necessary maintenance requirements.

Automatic blanket wash systems

Solvent

15 The solvent used in an automatic blanket wash system should be the same as that used when the system was commissioned (see below). A different solvent should not be used without reference to the supplier of the automatic blanket wash system. Recommissioning may be necessary. Solvent containers should identify clearly the solvent concerned.

16 The blanket wash supplier should ensure that the LFL at typical dryer operating temperatures (160-260°C) is available for the solvent used. If necessary arrangements should be made for the LFL to be determined experimentally.

17 The quantity of solvent applied onto the blanket cylinders should be metered and should be the minimum necessary. The quantity should be set during commissioning and any subsequent unauthorised access to the controlling mechanism should be prevented. Users should ensure that manual application of additional solvent is strictly prohibited. Where blanket wash drip trays are used, care should be taken to ensure that they do not overflow and allow the web to take excess solvent into the dryer.

Interface with press and dryer controls

18 Automatic means should be provided so that any mode of operation of the blanket wash system, eg on the fly, slow speed/end wash or pre-wet

* Note that percentage LFLs for ink and blanket wash solvents are additive.

start, can occur only under predetermined conditions whereby concentrations of solvent vapour within the dryer do not exceed 25% of the LFL. Such means include safety interlocks and systems to ensure that:

(a) the automatic blanket wash system cannot operate unless adequate exhaust ventilation of the dryer has been proved and the press speed is within the range established as safe during commissioning;

(b) the ink feed is interrupted and the inking rollers are lifted or the printing units are off impression before and during any blanket wash cycle;

(c) the maximum number of print units that can be washed at any one time (as established during commissioning) is not exceeded;

(d) slow speed/end washing can take place only with the web removed; and

(e) pre-wet start can be undertaken only with reduced quantities of solvent (compared with blanket washing) and at press speeds that are within a range established as safe during commissioning.

19 Where necessary to maintain solvent concentrations within the dryer below 25% of the LFL, an automatic time delay should operate after completion of one blanket wash cycle and before a further cycle can be commenced.

20 Where dryer exhausts from two or more press lines are ducted into the same incinerator or afterburner, automatic means should be provided to prevent any overloading of the pollution control equipment. This can be achieved by arranging that blanket washing cannot be in progress on more than one press line at a time.

Commissioning

21 Automatic blanket wash systems should not be brought into use until each installation has been thoroughly commissioned and safe operational limits established for all relevant parameters of the dryer, blanket wash system and press during any available modes of operation.

22 Commissioning should incorporate direct measurement of the maximum concentration of solvent within the dryer during the blanket wash cycle using, for example, a flame ionisation detector calibrated for the solvent concerned. Measurements should be taken both above and below the web, starting with minimum solvent application on one print unit. The quantity should be gradually increased while monitoring continuously the solvent levels in the dryer. Additional print units should be added sequentially.

23 Commissioning tests should be done using the maximum web width and should include all types of paper likely to be used, as lower quality or uncoated papers may lead to higher solvent levels inside the dryer. Before using paper outside the range originally commissioned, the user should inform the supplier of the blanket wash system and arrangements should be made for any necessary recommissioning. Recommissioning may also be necessary following addition or replacement of plant, eg installation of afterburners or other pollution control equipment.

24 Measurement of solvent levels in the dryer should be repeated at least annually and appropriate records kept.

Information and training

25 The supplier of the blanket wash system should provide users with adequate written information about the conditions under which the system has been commissioned, the limitations under which the system can be used, the safety devices incorporated, the results of commissioning and the necessary maintenance requirements.

26 Managers, operators and assistants should be adequately trained in the explosion hazard associated with blanket washing and in the necessary precautionary measures. In particular they should understand the significance of LFLs and the limitations under which blanket washing is to be undertaken. Procedures for blanket washing should be set out in written safe systems of work. Health and safety data sheets should be provided by the suppliers of the blanket wash solvents used.

REFERENCES

1 *Guidance from the Printing Industry Advisory Committee* IAC/L26*

2 *Printers - are you complying with COSHH?* IAC/L42*

3 *Providing for occupational health in the printing industry* IAC/L41*

4 *Fire Precautions Act 1971* chapter 40 HMSO ISBN 0 10 544071 X

5 *Health and Safety at Work etc Act 1974* chapter 37 HMSO ISBN 0 10 543774 3

6 *Guide to fire precautions in existing places of work that require a fire certificate* Factories, Offices, Shops and Railway Premises. Home Office and Scottish Home and Health Department HMSO ISBN 0 11 340906 0

7 *Code of practice for fire precautions in factories, offices, shops and railway premises not required to have a fire certificate* Home Office and Scottish Home and Health Department HMSO ISBN 0 11 340904 4

8 *Fire safety at work* Home Office and Scottish Home and Health Department HMSO ISBN 0 11 340905 2

9 *Guide to the Health and Safety at Work etc Act 1974* L1 HMSO ISBN 0 11 885555 7

10 *Articles and substances used at work* IND(G)1(L) (Rev)*

11 *COSHH and section 6 of the Health and Safety at Work Act* IND(G)97(L)*

12 *Substances for use at work: the provision of information* HS(G)27 2nd edition HMSO ISBN 0 11 885458 5

13 *Chemicals in the printing industry: the provision of health and safety information by manufacturers, importers and suppliers of chemical products to the printing industry* HMSO ISBN 0 11 883852 0

14 *Assessment of fire hazards from solid materials and the precautions required for their safe storage and use* HS(G) 64 HMSO ISBN 0 11 885654 5

15 *Reporting an injury or a dangerous occurrence* HSE 11 (Rev)*

16 *Notification and marking of sites. The Dangerous Substances (Notification and Marking of Sites) Regulations 1990* HS(R)29 HMSO ISBN 0 11 885435 6

17 *A guide to the Classification, Packaging and Labelling of Dangerous Substances Regulations 1984* HS(R)22 HMSO ISBN 0 11 883794 X

18 *A guide to the Control of Industrial Major Accident Hazards Regulations 1984* HS(R)21 (Rev) HMSO ISBN 0 11 885579 4

19 *A guide to the Notification of Installations Handling Hazardous Substances Regulations 1982* HS(R)16 HMSO ISBN 0 11 883675 7

20 *Fire Certificates (Special Premises) Regulations 1976* SI 1976/2003 HMSO ISBN 0 11 062003 8

21 BS 5588: Part 8: 1988 *Fire precautions in the design and construction of buildings - Code of practice for means of escape for disabled people*

22 *Writing a safety policy statement - Advice to employers* HSC 6*

23 *Making your safety policy work* IAC L/14*

24 *Control of Substances Hazardous to Health Regulations 1988* SI 1988/1657 HMSO ISBN 0 11 087657 1

25 *Control of substances hazardous to health* and *Control of carcinogenic substances* Approved Codes of Practice L5 HMSO ISBN 0 11 885698 7

26 *Training for health and safety in the printing industry* IAC/L66*

27 *Monitoring for health and safety in print - a guide to management action* IAC/L65*

28 *Code of practice for occupational fire brigades* Loss Prevention Council, 140 Aldersgate Street, London EC1A 4HY

29 BS5839: Part 1: 1988 *Fire detection and alarm systems for buildings - Code of practice for system design, installation and servicing*

30 *Gaseous fire extinguishing systems: precautions for toxic and asphyxiating hazards* Guidance Note GS 16 HMSO ISBN 0 11 883574 2

31 BS 5306 *Fire extinguishing installations and equipment on premises* (in various parts and sections)

32 BS 5423: 1987 *Specification for portable fire extinguishers*

33 BS 4547: 1972 *Classification of fires*

34 *Guidance on permit-to-work systems in the petroleum industry* (Revised edition) HMSO ISBN 0 11 885688 X

35 *Hot work: welding and cutting on plant containing flammable materials* HS(G)5 HMSO ISBN 0 11 883229 8

36 *The cleaning and gas freeing of tanks containing flammable residues* Guidance Note CS15 HMSO ISBN 0 11 883518 1

37 *Safe use and storage of flexible polyurethane foam* HS(G)1 HMSO ISBN 0 11 883208 5

38 *Storage of packaged dangerous substances* HS(G) 71 HMSO ISBN 0 11 885989 7

39 *Standard for the storage of roll paper* ANSI/ NFPA 231F - 1987 National Fire Protection Association, American National Standards Institute, New York

40 *The Loss Prevention Council Rules for automatic sprinkler installations* Loss Prevention Council, 140 Aldersgate Street, London EC1A 4HY

41 *The keeping of LPG in cylinders and similar containers* Guidance Note CS4 HMSO ISBN 0 11 883539 4

42 *Storage of LPG at fixed installations* HS(G)34 HMSO ISBN 0 11 883908 X

43 BCGA Code of Practice CP6 *The safe use of acetylene in the pressure range 0-1.5 bar (0-22 LBF per square inch)* British Compressed Gases Association, 26 Brighton Road, Crawley, West Sussex RH10 6AA

44 BCGA Code of Practice CP7 *The safe use of oxy-fuel gas equipment* British Compressed Gases Association

45 BCGA Code of Practice CP8 *The safe storage of gaseous hydrogen in seamless cylinders and similar containers* British Compressed Gases Association

46 *Oxygen - fire and explosion hazards in the use and misuse of oxygen* HSE 8 (Rev)*

47 *The safe use of flammable liquids* HSE Guidance Note (in preparation)

48 *Evaporating and other ovens* (under revision) HS(G)16 HMSO ISBN 0 11 883433 9

49 *Flame arrestors and explosion reliefs* HS(G)11 HMSO ISBN 0 11 883258 1

50 BS 5345: *Code of practice for selection, installation and maintenance of electrical apparatus for use in potentially explosive atmospheres: (other than mining applications or explosives processing and manufacture)* Part 1: 1989 *General recommendations* and Part 2: 1983 *Classification of hazardous areas*

51 *Electrical apparatus for use in potentially explosive atmospheres* HS(G)22 HMSO ISBN 0 11 883746 X

52 *Highly flammable liquids in the paint industry* HS(G) 4 HMSO ISBN 0 11 883219 0

53 *Electricity and flammable substances - a short guide for small businesses* Institution of Chemical Engineers, 165-171 Railway Terrace, Rugby, Warwickshire CV21 3HQ ISBN 085 295250 3

54 *Diesel-engined lift trucks in hazardous areas* Guidance Note PM 58 HMSO ISBN 0 11 883535 1

55 *The storage of flammable liquids in fixed tanks (up to 10 000 cubic metres total capacity)* HS(G)50 HMSO ISBN 0 11 885532 8

56 *The storage of flammable liquids in containers* HS(G)51 HMSO ISBN 0 11 885533 6

57 *The storage of flammable liquids in fixed tanks (exceeding 10 000 cubic metres total capacity)* HS(G)52 HMSO ISBN 0 11 885538 7

58 *Magnesium (Grinding of Castings and Other Articles) Special Regulations 1946* SR and O 1946/2107 HMSO

59 *The safe handling and storage of aluminium powder and paste* Aluminium Federation, Broadway House, Calthorpe Road, Five Ways, Birmingham B15 1TN

60 *Guide to dust explosion, prevention and protection* (in 3 parts) Institution of Chemical Engineers, 165-171 Railway Terrace, Rugby, Warwickshire CV21 3HQ

Schofield C *Guide to dust explosion, prevention and protection* Part 1: *Venting* Institution of Chemical Engineers 1984 ISBN 0 85 295177 9

Schofield C and Abbott J A *Guide to dust explosion, prevention and protection* Part 2: *Ignition prevention, containment, inerting, suppression and isolation* Institution of Chemical Engineers 1988 ISBN 0 85 294222 8

Lunn G A *Guide to dust explosion, prevention and protection* Part 3: *Venting of weak explosions and the effect of vent ducts: a British materials handling board design guide for practical systems* Institution of Chemical Engineers 1988 ISBN 0 85 295230 9

61 *Chemical plant and petroleum refinery piping* (with addenda) ANSI/ASME Standard B31.3-1990 American National Standards Institute. Engineering Equipment and Material Users Association *Supplement* to ANSI/ASME Standard B31.3 EEMUA London

62 BS 5958: Part 2: 1988 *Code of practice for control of undesirable static electricity: Recommendations for particular industrial situations*

63 *Electrostatic ignition - hazards of insulating materials* HSE Occasional Paper Series OP 5 HMSO ISBN 0 11 883629 3

64 *Generation and control of static electricity where flammable liquids are stored and used* Paintmakers Association of Great Britain Limited, Alembic House, 93 Albert Embankment, London SE1 7TY

65 BS 5145: 1989 *Specification for lined industrial vulcanized rubber boots*

66 BS 7193: 1989 *Lined lightweight rubber overshoes and overboots*

67 BS 1870 *Safety footwear* (in 3 parts)

68 BS 2050: 1978 *Specification for electrical resistance of conducting and antistatic products made from flexible polymeric material*

69 *Industrial use of flammable gas detectors* Guidance Note CS 1 HMSO ISBN 0 11 883948 9

70 BS 5274: 1985 *Specification for fire hose reels (water) for fixed installations*

71 BS 6575: 1985 *Specification for fire blankets*

72 BS 476 *Fire tests on building materials and structures* (in various parts and sections)

73 *Solvents and you* IND(G)93(L)*

74 BS 5885 *Automatic gas burners* (in 2 parts)

75 *Code of practice for the use of gas in low temperature plant* (2nd edition) British Gas Publication IM18 British Gas 1988

76 *Venting of deflagrations* ANSI/NFPA 68 National Fire Protection Association, Quincy, Mass, 1987

77 Marshall A R 'Gaseous and dust explosion venting: determination of explosion relief requirements' in *Loss prevention and safety promotion in the process industries* 3rd International Symposium, Basle 15-19 Sept 1980. European Federation of Chemical Engineering, Swiss Society of Chemical Industries 1980

British Standards are available from: British Standards Institution, Sales Department, Linford Wood, Milton Keynes MK14 6LE

* Leaflets available free from HSE area offices or information centre at:

Broad Lane, Sheffield S3 7HQ (Tel: 0742 892346)

PRINTING INDUSTRY ADVISORY COMMITTEE PUBLICATIONS

Safety in newspaper production HMSO
ISBN 0 11 883677 3

Safety in the use of inks, varnishes and lacquers cured by ultra-violet light HMSO
ISBN 0 11 885506 9

Ink fly in newspaper pressrooms HMSO
ISBN 0 11 883751 6

Safety in the use of isocyanate pre-polymers in the printing and printed packaging industries HMSO
ISBN 0 11 883847 4

Chemicals in the printing industry: the provision of health and safety information by manufacturers, importers and suppliers of chemical products to the printing industry HMSO ISBN 0 11 883852 0

Making your safety policy work: the implementation of safety policies in printing works IAC/L14*

Noise reduction at buckle folding machines HMSO
ISBN 0 11 883849 0

Health and safety for small firms in the print industry HMSO ISBN 0 11 883851 2

The control of lead in the printing industry IAC/L24*

Safety in the use of chemical products in the printing industry HMSO ISBN 0 11 883956 X

Control of health hazards in screen printing HMSO
ISBN 0 11 883973 X

Safe handling of materials in the printing industry HMSO ISBN 0 11 883983 7

Noise reduction at web-fed presses HMSO
ISBN 0 11 883972 1

Safety at power operated paper cutting guillotines HMSO ISBN 0 11 885460 7

Precautions against humidifier fever in the print industry IAC/L28*

Providing for occupational health in the printing industry IAC/L41*

Printers - are you complying with COSHH? IAC/L42*

Laser safety in printing HMSO
ISBN 0 11 885436 4

Monitoring for health and safety in print - a guide to management action IAC/L65*

Training for health and safety in the printing industry IAC/L66*

* Leaflets available free from HSE area offices or information centre at:

Broad Lane
Sheffield
S3 7HQ
Tel: 0742 892346

Printed in the United Kingdom for HSE, published by HMSO C50 10/92